The Polyol Paradigm and Complications of Diabetes

Margo Panush Cohen

The Polyol Paradigm and Complications of Diabetes

With a Foreword by Harold Rifkin

With 25 Figures

Springer-Verlag
New York Berlin Heidelberg
London Paris Tokyo

Margo Panush Cohen, M.D.,Ph.D.
Professor of Medicine and Director
Division of Endocrinology and Metabolism
University of Medicine and Dentistry of New Jersey
Newark, New Jersey 07103
USA

The quotation on pp. 2–3 is reprinted by permission of *The Wall Street Journal*, © Dow Jones and Company, Inc. 1986. All rights reserved.

Library of Congress Cataloging in Publication Data
Cohen, Margo P.
   The polyol paradigm and complications of diabetes.
   Bibliography: p.
   Includes index.
   1. Diabetes—Complications and sequelae.
2. Polyols—Physiological effect. 3. Aldose
reductase. I. Title. [DNLM: 1. Diabetes Mellitus—
complications. 2. Diabetes Mellitus—enzymology.
3. Sugar Alcohol Dehydrogenases—metabolism.
WK 895 C678p]
RC660.C474 1987   616.4'62   86-26022

Typeset by Publishers Service, Bozeman, Montana.

9 8 7 6 5 4 3 2 1

ISBN-13: 978-1-4612-9108-4   e-ISBN-13: 978-1-4612-4670-1
DOI: 10.1007/978-1-4612-4670-1

To Louis and Tillie Panush

and to Perry, Michael, Daniel, and Jonathan Cohen

Foreword

In the last decade, it has become increasingly evident that the clinical and morphologic changes underlying many of the complications of diabetes, including cataract formation, retinopathy, nephropathy, neuropathy, and macrovascular disease, are preceded by a variety of disturbances of biochemical and physiologic origin. Dr. Cohen has recently written a superb monograph, entitled *Diabetes and Protein Glycosylation: Measurement and Biologic Relevance*, in which she thoroughly explores how enhanced nonenzymatic glycosylation in uncontrolled diabetes underscores the pressing need for maintenance of long-term euglycemia. In the present volume, *The Polyol Paradigm and Complications of Diabetes*, she reviews, in a most succinct and thorough manner, how another biochemical mechanism, involving the polyol pathway, is involved in the pathogenesis of such diabetes complications as retinopathy, neuropathy, nephropathy, and cataract formation.

Dr. Cohen gives us a clearly written and comprehensive monograph, reviewing the chemistry of the polyol pathway and of the aldose reductase inhibitors, and the pathophysiologic significance of increased polyol pathway activity in a variety of tissues affected by diabetes mellitus. She insightfully describes the relationship of increased polyol pathway activity to altered metabolism of inositol-containing phospholipids and to changes in various tissue concentrations of *myo*-inositol. Finally, she provides us with a careful review of the existing experimental and clinical studies with a variety of different aldose reductase inhibitors that have been and are being performed in the hope of preventing or reversing long-term complications of diabetes.

This monograph is a most welcome addition to the literature on diabetes and its chronic complications, particularly since its author is a world-renowned endocrinologist and diabetes specialist who, throughout the years, has focused the major portion of her research activities on the biochemical and metabolic aberrations underlying the long-term complications of diabetes. It should appeal to the basic and clinical investigator, as well as to the clinician, and should be present in every medical school and hospital library where it is easily accessible and available to all who are concerned and involved with the management of diabetic patients.

HAROLD RIFKIN, M.D.

Clinical Professor of Medicine
Albert Einstein College of Medicine

Professor of Clinical Medicine
New York University School of Medicine

Principal Consultant
Diabetes Research and Training Center
Albert Einstein College of Medicine
Montefiore Medical Center
New York, New York

Preface

The polyol paradigm is one of two major theories that have been advanced to explain the pathogenesis of the complications of diabetes. The other, which implicates excess nonenzymatic glycosylation, has been described in the companion volume *Diabetes and Protein Glycosylation*, also published by Springer-Verlag.

The polyol pathway originally was described in the ocular lens, where it was implicated in the development of sugar cataracts. It now is clear that the polyol pathway is present in diverse tissues, where it may participate in the pathogenesis of diabetic neuropathy, retinopathy, and microvascular disease.

The pathway involves two enzymes, aldose reductase and sorbitol dehydrogenase, the first of which converts sugars into their respective alcohols (polyols). It is believed that activation of the polyol pathway has deleterious metabolic effects in several organs. Blocking the pathway by inhibiting aldose reductase, the responsible enzyme, thus may interrupt or forestall development of tissue damage in diabetes.

The Polyol Paradigm and Complications of Diabetes comprehensively reviews the biochemistry and pathophysiologic consequences of enhanced polyol pathway activity, with discussions oriented around the organ systems affected. Insights gained from experimental studies and clinical trials with aldose reductase inhibitors are critically analyzed.

M.P. COHEN

Contents

CHAPTER 1

Introduction

The polyol paradigm is one of two major theories that have been advanced to explain complications of diabetes. The other theory concerns the nonenzymatic glycosylation of proteins,[1] a process that is increased in the presence of hyperglycemia. Like excess nonenzymatic glycosylation, activation of the polyol pathway occurs in tissues that do not require insulin for glucose transport and can dispose of glucose via insulin-independent pathways. It is of more than passing interest that such tissues are the sites of several characteristic complications of diabetes, a fact that increases the attractiveness of the concept that enhanced flux of glucose through certain biochemical pathways in cells that are freely permeable to glucose is a mechanism underlying the development of chronic complications of diabetes.

The polyol pathway involves two enzymes, aldose reductase and sorbitol dehydrogenase, the first of which converts sugars into their respective alcohols (polyols). Initially reported to be operative in the development of sugar cataracts in the ocular lens, the polyol pathway, it is now clear, is present in diverse tissues, where it may participate in the pathogenesis of diabetic neuropathy, retinopathy, and microvascular disease.

The concept that tissue accumulation of polyol pathway products, or some consequence thereof, has deleterious effects in involved organs has gained credence in recent years, as has the notion that blocking the pathway by inhibiting the responsible enzyme could interrupt or forestall the evolution of tissue damage in patients with diabetes. Indeed, these ideas have attracted the attention not only of physicians and scientists concerned with the treatment, prevention, and biochemical basis of diabetic complications, but also of patients, the pharmaceutical industry, drug analysts, and investment houses. Excerpts from a feature article that appeared in the *Wall Street Journal*[2] attest to the interest that has been generated in the pharmaceutical and financial communities and among the lay public regarding the polyol pathway and aldose reductase inhibitors:

In 1966, Mr. Dvornik, then chief biochemist at Ayerst Laboratories, listened as two Harvard University scientists described the previously unknown chain of biochemical events through which diabetes ravages the body. Disrupt that chain with a synthetic chemical, the two scientists reasoned, and Ayerst would have the first drug for treating the complications that can cripple, and sometimes kill, diabetics.

Today, Ayerst, a unit of New York-based American Home Products Corp., is one of six companies racing to the market with drugs based on the two scientists' idea. Although the anti-diabetes drugs may still be years away from pharmacy shelves, expectations for them are high. The drugs, called aldose reductase inhibitors (ARI), "may revolutionize the treatment of diabetes". . . .

The ARI drugs are among the first fruits of the so-called rational approach to drug research that has been the source of much optimism in the industry. . . .

The development of the drug, however, . . . has been slowed by difficult and expensive testing and by competition with other drug companies to come up with the most effective product. Almost every pharmaceutical company is pursuing ARI research because it's based on such a clearly defined scientific approach. And with five million diabetics in America, the potential for profit is high.

Diabetes is caused by a shutdown in production of insulin. Without insulin, muscle and fat cells can't process the glucose, or sugar, that they use as fuel. High levels of unused sugar back up in the bloodstream, then nerve, eye and kidney cells, which don't need insulin to process the glucose, gorge themselves on the excess sugar.

Scientists now believe that years of such overconsumption eventually damages the eye's cells, causing cataracts and even blindness; disrupts nerve cell activity, causing severe pain or loss of feeling; and upsets kidney cells, sometimes so severely that the organ stops functioning.

But scientists were powerless to stop the process until they found a chemical switch to close the pathway involved in the cells' sugar feast. As with many of the diseases that have been approached rationally, the key to the diabetes pathway was an enzyme. And it wasn't until 1969 that chemists figured out how to interrupt the activities of enzymes, which are enormously powerful proteins that act as catalysts and speed up biochemical reactions.

The enzyme that unlocked the mystery of the diabetes pathway was noticed in 1963 by Jin Kinoshita. . . . "We figured we'd locked into the common pathway through which diabetes affected all organs," says Dr. Kinoshita. . . "The next moves seemed relatively simple: Inhibit the enzyme, block the pathway, and disrupt the disease". . . .

Ayerst, which had believed itself to be alone in the ARI field, received a shock: Pfizer, Inc., which had been doing its own ARI research, had developed a drug called Sorbinil that was as effective as Ayerst's but was 25 times more potent. . . .

But proving the drug's effectiveness has been a slow process. Recent studies published in medical journals show that the drug's ability to reduce nerve damage in longtime diabetes is limited. Doctors who have worked with ARIs believe that "once the damage has been done to the cells, nothing can truly reverse it". . . . Meanwhile, Dr. Gabbay is worried that as a preventative, the drugs will be used to block the action of the aldose reductase enzyme for many years. "Since we still don't know what the enzyme's job in healthy tissue is, blocking the enzyme over the course of a lifetime may be very dangerous."

This book is written in the belief that cognizance of the potential pathophysiologic consequences of activation of the polyol pathway by hyperglycemia will reinforce the conviction that every effort should be made to achieve and maintain normalization of blood glucose levels in diabetic patients. Like its recently published companion volume, *Diabetes and Protein Glycosylation*,[1] *The Polyol Paradigm* adds another dimension to the concept that the guilt of glucose in the pathogenesis of complications of diabetes can no longer be denied. The chapters that follow review the chemistry of the polyol pathway and of aldose reductase inhibitors, discuss the metabolic and pathophysiologic significance of enhanced polyol pathway activity, and analyze the insights gained from experimental and clinical studies with aldose reductase inhibitors. Although the ultimate role of the polyol pathway in the sequence of events leading to several of the complications of diabetes, and of aldose reductase inhibitors in the treatment or prevention of these complications, remains to be determined, it is believed that comprehensive dis-

cussion of this topic is of timely importance to all concerned with
the management of diabetic patients and interested in the mechanisms responsible for the complications characteristically associated with diabetes.

References

1. Cohen MP: *Diabetes and Protein Glycosylation: Measurement and Biologic
   Relevance*. New York, Springer-Verlag, 1986.
2. Waldholz M: Researchers use logic to uncover, then break, the chain of diabetes.
   The Wall Street Journal, February 5, 1986, p 31.

Chemistry

Aldose Reductase and Sorbitol Dehydrogenase

Aldose reductase is a member of the aldo-keto reductase enzyme family that is present in mammalian tissues. It is a monomeric NADPH-binding protein that catalyzes the reduction of aldoses to their corresponding sugar alcohols. Using NADPH as coenzyme, aldose reductase catalyzes the initial reaction of the sorbitol pathway, also called the polyol pathway.[1-3] Oxidation of sorbitol to fructose, catalyzed by the enzyme sorbitol dehydrogenase, which uses the cofactor NAD as the electron acceptor, constitutes the second reaction in this pathway, according to the scheme depicted in Figure 2-1.

Both aldose reductase and sorbitol dehydrogenase have broad substrate specificities, including glucose and galactose for the former enzyme, and xylitol (to form D-xylulose) and ribitol (to form D-ribulose) for the latter. Notably, however, sorbitol dehydrogenase has limited ability to further metabolize galactitol, the sugar alcohol formed from galactose. This feature has been exploited experi-

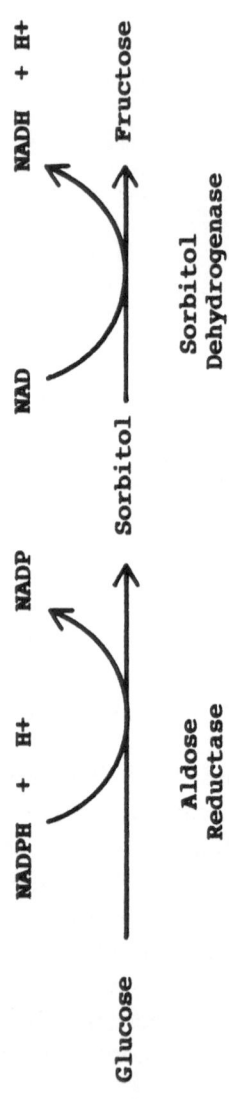

FIGURE 2-1 The sorbitol pathway: Aldose reductase (alditol-NADP+ oxidoreductase, EC 1.1.1.21) and Sorbitol dehydrogenase (L-iditol dehydrogenase) (EC 1.1.1.14).

mentally with the use of galactose-fed models to probe the consequences of polyol accumulation in various tissues.[4]

Because of their low affinity for glucose, the aldose reductases are normally operative at low catalytic rates when glucose concentration is in the physiologic range. In the presence of hyperglycemia or galactosemia, however, their activity increases substantially, causing increased formation of the correspondent sugar alcohols sorbitol and galactitol.[5,6] This fact, coupled with the recognition that sugar alcohols penetrate cell membranes poorly, and hence, once formed, are trapped intracellularly, led to the proposal that osmotic effects consequent to polyol accumulation are instrumental in the development of certain chronic complications of diabetes.[7-9] This theory is discussed more fully in later sections. Appreciation of the potential role of the polyol pathway in the pathogenesis of diabetic complications also generated considerable interest in the identification and characterization of aldose reductases in various tissues, and led to the search for effective inhibitors of aldose reductase that might prove clinically useful for the treatment or prevention of diabetic sequelae.

Biochemical studies have confirmed the presence of aldose reductase activity in diverse tissues, including seminal vesicles, placenta, lens, brain, erythrocyte, muscle, renal papillae, and aorta.[1,2,10-42] In early reports, the distinction between aldose reductase and other enzymes, such as aldehyde reductases or hexonate dehydrogenase, that have closely related substrate specificities was sometimes imperfect; however, immunologic studies have clearly established the separate identity of aldose reductase.[36,37,42,43] Immunohistochemistry also has demonstrated widespread distribution of aldose reductase in mammalian tissues[44-46] (Tables 2-1, 2-2). The enzyme is present in lower organisms such as ycast, which has provided a stable and convenient source of aldose reductase with high specific activity that has proved useful for studies of the reaction mechanism.[47,48] It is also found in the muscle of the parasitic nematode *Ascaris suum*.[49]

Aldose reductase is a member of a class of several pyridine nucleotide-linked enzymes which share common structural and functional characteristics and which catalyze the oxidation and reduction of aldehydes. Although it has broad substrate specificities which overlap those of aldehyde reductases, the enzyme has some

TABLE 2-1 Tissue Distribution of Aldose Reductase Detected by Immuno-histochemistry

Tissue	Localization
Adrenal cortex	Glomerulosa, fasciculata, reticularis
Cerebral cortex	Pyramidal cells, choroid plexus (epithelium)
Blood vessels	Intima (endothelium), media
Eye	Lens, retina, cornea, ciliary processes, optic nerve
Intestine	Lamina propria of jejunal villi
Kidney	Interstitial cells, collecting tubules, Henle's loop, papillae (epithelium)
Ovary	Medulla, oocyte, granulosa cord cells
Peripheral nerve	Schwann cells
Salivary glands	Ductal epithelium
Spinal cord	Anterior horn cell bodies, white matter neuroglia cells
Testis	*Leydig cells, Sertoli cells, epididymis

Data from Kern and Engerman.[44]
*According to Ludvigson et al.,[45] aldose reductase is specifically located in the Sertoli cell and is not detected immunocytochemically in any other testicular cell type.

affinity for D-glucose and other aldo-sugars, resulting in retention of its separate name and classification as an aldose reductase. However, the aldehyde form of D-glucose may be the true substrate in the reduction of D-glucose by the enzyme. Aldose reductase is an A-type enzyme, meaning that the reaction it catalyzes occurs with the direct and stereospecific addition and removal of hydrogen from the para position of the nicotinamide ring of the nucleotide.[15,21,50-52] Sorbitol dehydrogenase is also an A-type enzyme. Aldose reductase in the lens, the tissue that has been most intensively studied, of several species is a monomeric acidic enzyme having a molecular weight in the range of 35,000 to 40,000 and an isoelectric point of 4.75 to 4.85.[25,26,29,30] Human brain aldose reductase is a monomer of molecular weight 38,000 with an isoelectric point of 5.9.[21]

Human placental aldose reductase has an apparent molecular weight of 37,000 on SDS-gel electrophoresis.[14] On immunodiffusion or immunoelectrophoresis, antibodies raised against this preparation give a single line of identity with either human placental or human lens aldose reductase, but not with rat lens aldose reductase.

TABLE 2-2   Tissues in Which Immunohistochemical Evidence of Aldose Reductase Is Lacking

Adipose tissue
Adrenal medulla
Kidney cortex
Lung
Microvasculature
Oviduct
Parathyroid
Prostate
Salivary gland acini
Epidermis and dermis
Spleen
Thyroid
Umbilical cord
Uterus

Data from Kern and Engerman.[44]

Thus, lens aldose reductases from different species are not immunologically identical. On the other hand, antibodies against purified rat lens aldose reductase, which appears as a closely spaced doublet with a molecular weight of approximately 38,000 on SDS-gel electrophoresis, show cross-reactivity with human placental aldose reductase. Both enzymes are activated by sulfate ions and inhibited by chloride ions, and have similar substrate specificities with higher affinities for aromatic and aliphatic aldehydes than for hexose sugars, greater activity with aromatic than with aliphatic aldehydes, and greater preference for short-chain than for long-chain aliphatic aldehydes. In purified form, neither shows any activity when L-gulonate (gulonic acid) is used as substrate, thus differentiating them from L-hexonate dehydrogenase.

When isolated from brain, lens, and other tissues, aldose reductase is closely associated with an enzyme of the glucuronic acid-xylulose shunt, NADP-L-hexonate dehydrogenase (NADP+ 1-oxidoreductase; EC 1.1.1.19). In muscle, the properties of one of two major aldehyde reductases, which had been designated ARI and ARII (AR 1 and AR 2), appear to be similar to those of lens aldose reductase. This "low-K_m" aldehyde reductase in muscle has immunologic identity with similar low-K_m aldehyde reductases of brain and kidney and with lens aldose reductase.[37,38] These muscle and kidney low-K_m aldehyde reductases, and probably the high-K_m aldehyde reductases as well, are inhibited by the commercial aldose reductase inhibitors alrestatin and Sorbinil. Thus, it has been argued that a separate designation for aldose reductase is not necessary. However, with the impetus provided by the pharmaceutical industry and the emergence of aldose reductase inhibitors in the clinical arena, this designation is now too firmly entrenched to be easily supplanted by a revised nomenclature.

In some studies, bovine, rat, and human lens aldose reductase have exhibited unusual kinetic characteristics, with a deviation from classic Michaelis-Menten kinetics that consisted of a concave downward curvature in Lineweaver-Burk double reciprocal plots at higher substrate concentrations. This behavior may reflect the presence of two enzymes. Although aldose reductase was originally described as a single species, recent studies with rabbit lens have identified two bands of aldose reductase activity by anionic poly-acrylamide gel electrophoresis.[38] Purified rat lens aldose reductase also migrates as a closely spaced doublet, but one of the bands may represent a product of proteolysis. In the rabbit lens, these closely migrating bands represented monomeric enzymes, and had elec-trophoretic mobilities identical to those of two aldose reductases (aldose reductase 1 and aldose reductase 2) purified from skeletal muscle of male rabbits. Double immunodiffusion studies showed that antiserum to the muscle aldose reductase 1 cross-reacted with complete identity with muscle aldose reductase 2 and 1, and that antiserum to muscle aldose reductase 2, which almost completely blocks the activity of this enzyme, cross-reacted with complete identity with the lens aldose reductases. The molecular weights of the lens and muscle aldose reductase 1 and the muscle aldose reduc-tase 2 were respectively estimated at 40,200 and 41,500, based on a curve derived from the relative mobilities of standard proteins plotted against their molecular weights. However, on gel filtration the molecular weights of the two muscle aldose reductases were indistinguishable at about 34,000. Like the lens aldose reductases, the enzymes from rabbit skeletal muscle are inhibited by the com-mercial aldose reductase inhibitors alrestatin and Sorbinil and by the flavinoid quercetin. According to one report, pig lens aldose reductase, in contrast to enzyme in the lens of the above-mentioned species, does not show homotrophic cooperative effects and does not appear to contain isoenzymes.[27] On the other hand, Markus and co-workers found that aldose reductase was present as two or more isoenzymes in all mammalian lenses they studied (human, cat, dog, guinea pig, monkey, pig, rabbit, rat, and sheep) except those of mouse.[53]

Although the presence of aldose reductase in red cells has been argued, it appears that human erythrocytes contain an aldose reductase that has kinetic, immunologic, and structural properties similar to the lens enzyme.[36,42] Despite some early confusion with the hexonate dehydrogenase present in these cells, it appears that erythrocyte aldose reductase is a separate enzyme since antiserum raised against the former enzyme does not cross-react with the latter. Aldose reductase from human erythrocytes has been purified to homogeneity.[36] The molecular weight of this enzyme, based on Sephadex gel filtration and polyacrylamide gel electrophoresis, is 32,500. It has an isoelectric pH of 5.47, a pH optimum of 6.2, and a requirement for lithium sulfate (0.4 M) for full expression of its activity. The enzyme is inhibited by various aldose reductase inhibitors and, in general, has properties similar to those exhibited by aldose reductases from bovine lens and brain and from human brain.

Under physiologic conditions, red cell aldose reductase exists in activated and unactivated forms, and exhibits biphasic kinetics.[54,55] The activated form reduces glucose to sorbitol with about seven times greater efficiency than does the unactivated form. Activation can be accomplished by preincubation of the enzyme with glucose-6-phosphate, NADPH, and glucose, 10 μM each. Aldose reductase in other tissues may also exist in activated and unactivated forms (Figure 2-2),[56] which could account for the previously observed anomalous kinetics and would help explain the increased aldose reductase activity and consequent polyol accumulation associated with hyperglycemia. Interestingly, the activated enzyme is less susceptible to inhibition by aldose reductase inhibitors such as Sorbinil, alrestatin, and quercitrin (Figure 2-3). This may explain the need in vivo for higher plasma levels of inhibitor to accomplish lowering of the sorbitol concentration in the erythrocytes of diabetic subjects than are required in vitro under standard assay conditions. Similarly, aldose reductase prepared from normal human lens exhibits properties similar to native enzyme from other tissues,

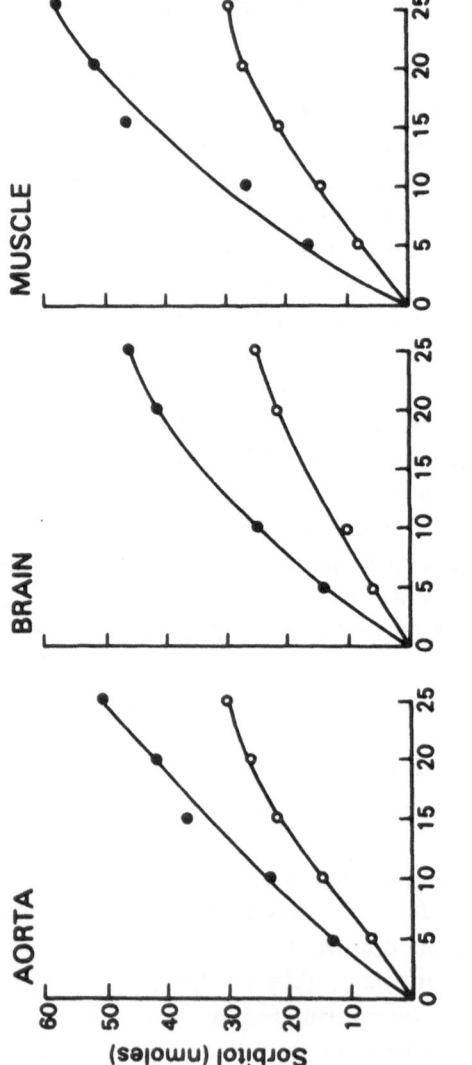

FIGURE 2-2 Formation of sorbitol by aldose reductase from various tissues. ○——○, unactivated enzyme; ●——●, enzyme activated by incubation with glucose-6-phosphate, NADPH, and glucose. Reproduced with permission from the American Diabetes Association, Inc. From Das B, Srivastava SK: Activation of aldose reductase from tissues. *Diabetes* 1985; 34:1145–1151.

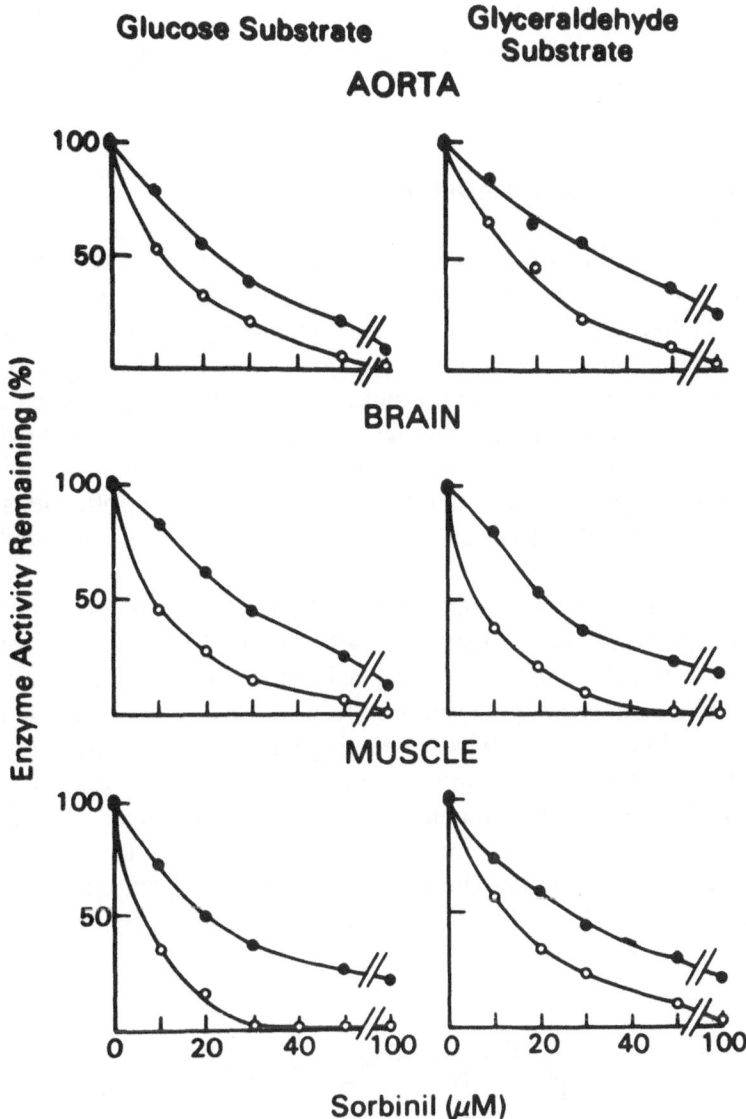

FIGURE 2-3   Inhibition by Sorbinil of activated (●——●) and unactivated (○——○) forms of aldose reductase from various tissues. Reproduced with permission from the American Diabetes Association, Inc. From Das B, Srivastava SK: Activation of aldose reductase from tissues. *Diabetes* 1985; 34:1145–1151.

whereas aldose reductase prepared from lens of hyperglycemic diabetic subjects exhibits properties similar to those of enzyme activated in vitro.

Aldose Reductase Inhibitors

The identification of aldose reductase in a variety of tissues in which complications of diabetes occur has led not only to a mechanistic theory, discussed in detail in the following chapters, proposing that hyperglycemia can alter the metabolism and function of cells in these tissues via sorbitol accumulation, but also to the search for inhibitors of aldose reductase that are nontoxic and effective in vivo with an appropriate duration of action. The hope is that such compounds will prove useful therapeutically for those complications of diabetes in which activation of the polyol pathway is believed to play a role.

A variety of structurally unrelated compounds possess aldose reductase-inhibiting properties. Early studies showing that long-chain fatty acids and α-keto fatty acids could inhibit aldose reductase when galactose, glucose, or xylose was used as substrate led to the development and in vitro testing of tetramethylene glutaric acid.[2,6,57-61] However, this compound did not prove useful in vivo because of its inability to penetrate plasma membranes.[62,63] Later, a number of heterocyclic compounds were found to be capable of inhibiting the enzyme. These include quinolones, spirohydantoins and structurally similar isoxazolidines, xanthones, flavinoids and

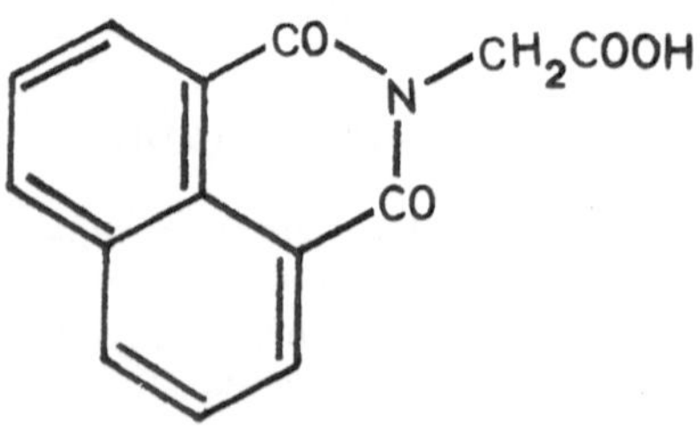

FIGURE 2-4 Structure of alrestatin (1,3-dioxo-1H-benz[de]isoquinolone-2(3H)acetic acid).

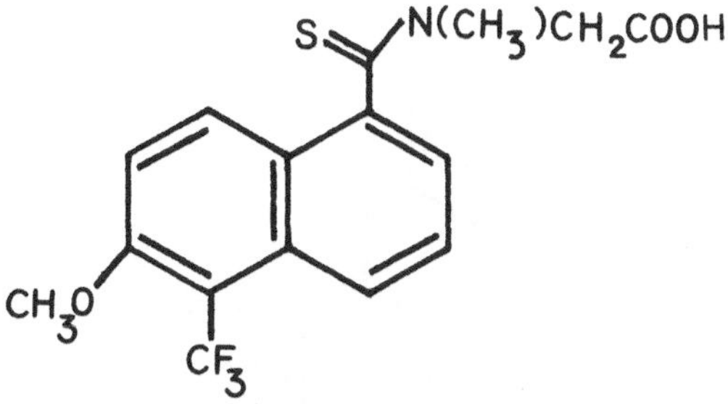

FIGURE 2-5   Structure of tolrestat (Alredase®; N-[5-trifluoromethyl)-6-methoxy-1-naphthalenyl]thioxomethyl]-N-methylglycine).

structurally related quercitrins, and numerous other compounds that contain a chromone ring system.[64-87] Several of these various compounds, such as alrestatin (Figure 2-4), tolrestat (Figure 2-5), Sorbinil (Figure 2-6), Clinoril® (Figure 2-7), Syntex (Figure 2-8), quercitrin (Figure 2-9), and antirheumatic drugs such as salicylates, indomethacin, and oxyphenbutazone, are now well known. Others such as Statil (ICI 10,5552; 1-[3,4-dichlorobenzyl]-3-methyl-1,2-dihydro-oxiquinol-4-ylacetic acid) and ONO 2235 ((E-3-carboxymethyl-5)[(2E)-methyl-3-phenylpropenylidene]rhodanine) have begun to appear in the literature more recently.

Kinetic studies show that most of these inhibitors act by noncompetitive or uncompetitive inhibition, indicating that they do not compete with either the substrate or the nucleotide cofactor site (Refs. 67,70,77,78,85,88–90). In fact, competition studies suggest

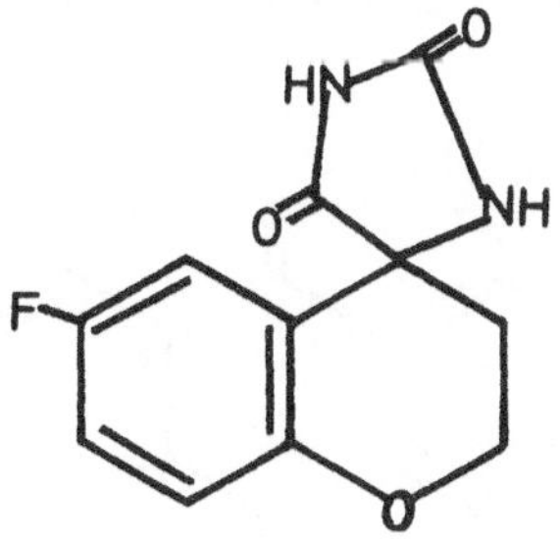

FIGURE 2-6   Structure of Sorbinil (d-6-fluoro-spiro(chroman-4-4'-imidazolidine)-2',5'-dione).

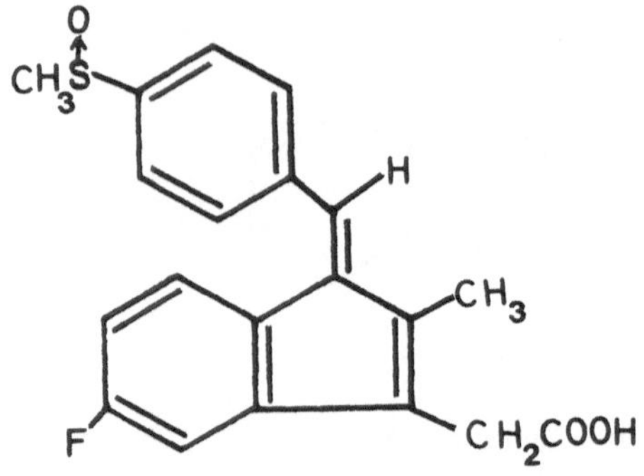

FIGURE 2-7   Structure of sulindac (Clinoril®; 2-(5-fluoro-2-methyl-1-[p-methyl-sulfinyl)benzylidine]indene-3-acetic acid).

that compounds with aldose reductase-inhibiting properties can interact reversibly at a common site, called the inhibitor site, on the aldose reductase enzyme. Inhibitors may act to convert the active enzyme to inactive forms, whereas maintenance of activity appears to require protection by thiols.[57,91] Moreover, it appears that aldose reductase inhibitors can also act as antioxidants, and in this manner may decrease tissue levels of toxic dicarbonyl compounds and free radicals produced spontaneously from auto-oxidation of monosaccharides.[83,92,93] In view of the superoxide theory of cataractogenesis, discussed in the chapter on aldose reductase and complications of the eye, this property must be taken into account in any interpretation of the beneficial effects of aldose reductase inhibitors on the development of sugar cataracts.

The inhibitor site of aldose reductase enzymes appears to be structurally distinct from those sites that bind either the substrate or the

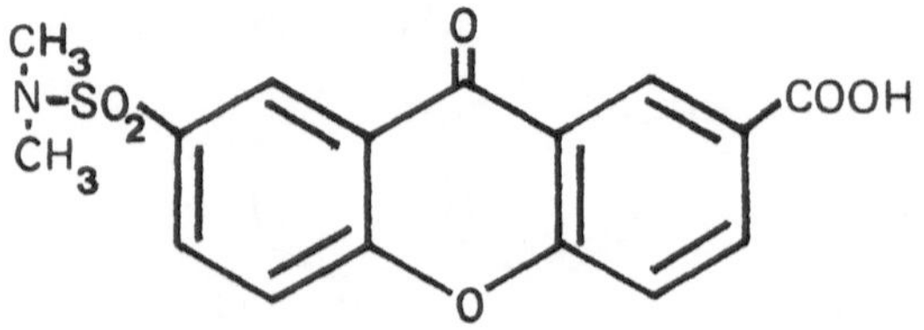

FIGURE 2-8   Structure of Syntex (7-dimethyl sulfamoyl-xanthone-2-carboxylic acid).

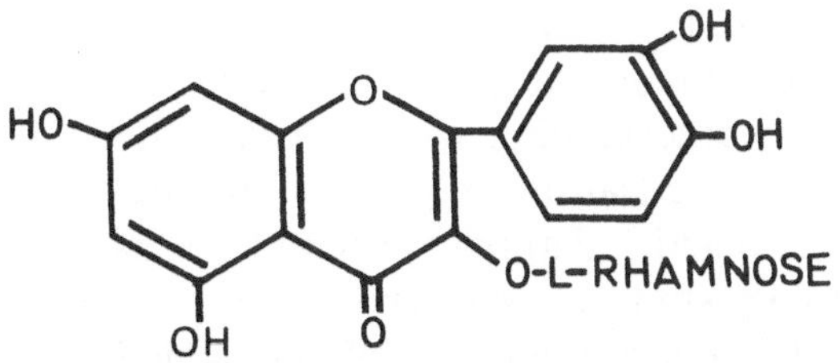

FIGURE 2-9 Structure of quercitrin (2-(3,4-dihydroxyphenyl)-3-O-rhamnosyl-5,7-dihydroxy-4-oxo-4H-chromen).

nucleotide cofactor. It can distinguish between various inhibitors stereochemically, recognizing stereospecific differences in inhibitor molecules.[88] The inhibitory property of various agents depends on the presence of components that are subject to nucleophilic attack. For compounds containing the benzopyran ring system, inhibitory activity has been correlated with the ability to undergo a charge-transfer interaction at a reactive carbonyl group by accepting a pair of electrons from the enzyme.[78] According to Kador and Sharpless, the inhibition of aldose reductase by diverse compounds could result from several mechanisms.[90] For example, binding of the inhibitor could alter the three-dimensional configuration of the enzyme, which could in turn lead to a hindrance of the catalytic site, either by conformational distortion or by steric interference produced by overlapping of the bound inhibitor. These possibilities could arise directly from nucleophilic attack of the reactive carbonyl to form a reversible tetrahedral intermediate, or from a charge-transfer bridge formed between the nucleophilic residue on the inhibitor site and an acidic moiety on another region of the protein. In the latter situation, the reactive carbonyl would be acting to allow nonproximate amino acid residues in the aldose reductase protein to interact. Tyrosine and arginine have been proposed as the respective acidic and basic residues involved in such an interaction. This would fit with the proposal that arginine, possibly together with lysine and histidine, is part of a basic center necessary for activity of the enzyme.[33,91]

These investigators have further postulated that enzyme inhibitor interactions involve a three-point site, and have offered a model for the inhibitor site based on information gained from computer molecular modeling, molecular orbital calculations, structure-activity relationships, and protein modifications. This model includes hydrophobic binding with two lipophilic regions on the enzyme and a charge-transfer interaction between the nucleophilic residue and the reactive carbonyl group of the inhibitor.

The foregoing studies were undertaken to define minimum requirements for aldose reductase inhibitory activity, and thereby provide a framework for the design of new agents with greater potency and specificity. This approach promises replacement of random screening techniques by development of rationally and specifically designed compounds. Such compounds could include agents that do not inhibit hexonate dehydrogenases, as many of the present aldose reductase inhibitors do.[67,94] For example, a recent study found that Sorbinil, alrestatin, and quercitrin are potent inhibitors of liver aldehyde reductase I and brain, liver, and erythrocyte aldehyde reductase II, as well as of purified lens and brain aldose reductase.[95] It could also allow development of inhibitors targeted for the aldose reductase of particular tissues, thus overcoming the differences in susceptibility to inhibition of the enzyme from different tissue sources that have been identified.[96-98] For example, the order of relative potency of four compounds against human placental aldose reductase was chromone derivatives > alrestatin > quercitrin > tetramethyleneglutaric acid, whereas the order of potency of these compounds against rat lens aldose reductase was quercitrin > alrestatin > chromone derivatives = tetramethylene glutaric acid.[96] These findings further suggested that evaluation of aldose reductase inhibitors for potential clinical use may require testing against human aldose reductase. Finally, the results of at least one study suggest that, in some tissues, aldose reductase might be associated in vivo with a factor that renders it insensitive to inhibitors.[99] This could have pharmacologic significance in the development and design of agents targeted for specific tissues.

References

1. Hers HG: Le mécanisme de la transformation de glucose en fructose par les vesicules séminales. *Biochim Biophys Acta* 1956; 22:202–203.
2. Hayman S, Kinoshita JH: Isolation and properties of lens aldose reductase. *J Biol Chem* 1965; 240:877–882.

3. Gabbay KH: The sorbitol pathway and complications of diabetes. *N Engl J Med* 1973; 288:831–836.

4. Cogan DG, Kinoshita JH, Kador PF, et al: Aldose reductase and complications of diabetes. *Ann Intern Med* 1984; 101:82–91.

5. Kinoshita JH, Futterman S, Satoh K, et al: Factors affecting the formation of sugar alcohols in ocular lens. *Biochim Biophys Acta* 1963; 74:340–350.

6. Chylack LT, Kinoshita JH: A biochemical evaluation of a cataract induced in a high glucose medium. *Invest Ophthalmol* 1969; 8:401–412.

7. LeFevre PG, Davies RI: Active transport into the human erythrocyte: Evidence from comparative kinetics and competition among monosaccharides. *J Gen Physiol* 1951; 34:515–524.

8. Wick AN, Drury DR: Action of insulin on the permeability of cells to sorbitol. *Am J Physiol* 1951; 166:421–423.

9. Kinoshita JH: Cataracts in galactosemia. *Invest Ophthalmol* 1965; 4:786–799.

10. Hers HG: Le mécanisme de la formation du fructose séminal et du fructose foetal. *Biochim Biophys Acta* 1960; 37:127–138.

11. Hastein T, Velle W: Placental aldose reductase activity and foetal blood fructose during bovine pregnancy. *J Reprod Fertil* 1968; 15:47–52.

12. Hastein T, Velle W: Purification and properties of aldose reductase from the placental and seminal vesicle of the sheep. *Biochim Biophys Acta* 1969; 178:1–10.

13. Clements R, Winegrad AI: Purification of alditol NADP oxidoreductase from human placenta. *Biochem Biophys Res Commun* 1972; 47:1473–1480.

14. Kador PF, Carper D, Kinoshita JH: Rapid purification of human placental aldose reductase. *Anal Biochem* 1981; 114:53–58.

15. Hoffman PL, Wermuth B, von Wartburg J-P: Human brain aldehyde reductases: Relationship to succinic seminaldehyde reductase and aldose reductase. *J Neurochem* 1980; 35:354–366.

16. Dons RF, Doughty CC: Isolation and characterization of aldose reductase from calf brain. *Biochim Biophys Acta* 1976; 452:1–12.

17. Boghosian RA, McGuiness ET: Affinity purification and properties of porcine brain aldose reductase. *Biochim Biophys Acta* 1979; 567:278–286.

18. Moonsammy GI, Stewart MA: Purification and properties of brain aldose reductase and L-hexonate dehydrogenase. *J Neurochem* 1967; 14:1187–1193.

19. O'Brien MM, Schofield PJ: Polyol pathway enzymes of human brain. *Biochem J* 1980; 187:21–30.

20. Boghosian RA, McGuiness ET: Pig brain aldose reductase: A kinetic study using centrifugal fast analyzer. *Int J Biochem* 1981; 13:909–914.

21. Wermuth B, Burgesser H, Bohren K, et al: Purification and characterization of human-brain aldose reductase. *Eur J Biochem* 1982; 127:279–284.

22. Ris MM, von Wartburg J: Heterogeneity of NADPH-dependent aldehyde reductase from human and rat brain. *Eur J Biochem* 1973; 37:69–77.

23. Turner AJ, Tipton KF: The characterization of two reduced nicotinamide-
    adenine dinucleotide phosphate-linked aldehyde reductases from pig brain.
    Biochem J 1972; 130:765–772.

24. Jedziniak JA, Chylack LT, Cheng H-M, et al: The sorbitol pathway in human
    lens: Aldose reductase and polyol dehydrogenase. *Invest Ophthalmol Vis Sci*
    1981; 20:314–326.

25. Sheaf CM, Doughty CC: Physical and kinetic properties of homogeneous
    bovine lens. *J Biol Chem* 1976; 251:2696–2702.

26. Hayman S, Lou MF, Merola LO, et al: Aldose reductase activity in the lens and
    other tissues. *Biochim Biophys Acta* 1966; 128:474–482.

27. Branlant G: Properties of an aldose reductase from pig lens. *Eur J Biochem*
    1982; 129:99–104.

28. Crabbe MJC, Halder AB: Affinity chromatography of bovine lens aldose
    reductase, and a comparison of some kinetic properties of the enzyme from
    lens and human erythrocyte. *Biochem Soc Trans* 1980; 8:194–195.

29. Conrad SM, Doughty CC: Comparative studies on aldose reductase from
    bovine, rat and human lens. *Biochim Biophys Acta* 1982; 708:348–357.

30. Hermann RK, Kador PF, Kinoshita JH: Rat lens aldose reductase: Rapid purifi-
    cation and comparison with human placental aldose reductase. *Exp Eye Res*
    1983; 37:467–474.

31. Inagaki K, Miwa I, Okuda J: Affinity purification and glucose specificity of
    aldose reductase in bovine lens. *Arch Biochem Biophys* 1979; 216:337–344.

32. Tanimoto T, Fukuda H, Kawamura J: Purification and some properties of
    aldose reductase from rabbit lens. *Chem Pharm Bull (Tokyo)* 1983;
    31:2395–2403.

33. Doughty CC, Lee S-M, Conrad S, et al: Kinetic mechanism and structural
    properties of lens aldose reductase, in Weiner H, Wermuth B (eds): *Enzymology
    of Carbonyl Metabolism: Aldehyde Dehydrogenase and Aldo/Keto Reductase*,
    pp 223–242. New York, Alan R. Liss, 1982.

34. Attwood MA, Doughty CC: Purification and properties of calf liver aldose
    reductase. *Biochim Biophys Acta* 1974; 370:358–368.

35. Halder AB, Wolff S, Ting H-H, et al: An aldose reductase from human erythro-
    cyte. *Biochem Soc Trans* 1980; 8:644–645.

36. Das B, Srivastava SK: Purification and properties of aldose reductase and alde-
    hyde reductase II from human erythrocyte. *Arch Biochem Biophys* 1985;
    238:670–679.

37. Cromlish JA, Flynn TG: Pig muscle aldehyde reductase. Identity of pig muscle
    aldehyde reductase with pig lens aldose reductase and with low Km aldehyde
    reductase of pig brain and pig kidney. *J Biol Chem* 1983; 258:3583–3586.

38. Cromlish JA, Flynn TG: Purification and characterization of two aldose reduc-
    tase isoenzymes from rabbit muscle. *J Biol Chem* 1983; 256:3416–3424.

39. Gabbay KH, O'Sullivan JB: The sorbitol pathway in diabetes and galactose-
    mia: Enzyme localization and changes in kidney. 1968; *Diabetes* 17:300.

40. Gabbay KH, Cathcart ES: Purification and immunologic identification of aldose reductases. *Diabetes* 1974; 23:460–468.

41. Morrison AD, Clements RS Jr, Winegrad AI: Effects of elevated glucose concentrations on the metabolism of the aortic wall. *J Clin Invest* 1972; 51:3114–3123.

42. Srivastava SK, Ansari NH, Hair GA, et al: Aldose and aldehyde reductase in human tissues. *Biochim Biophys Acta* 1984; 800:220–227.

43. Wirth H-P, Wermuth B: Immunochemical characterization of aldo-keto reductases from human tissues. *FEBS Lett* 1985; 187:280–282.

44. Kern TS, Engerman RL: Immunohistochemical distribution of aldose reductase. *Histochem J* 1982; 14:507–515.

45. Ludvigson MA, Sorenson RL: Immunohistochemical localization of aldose reductase. I. Enzyme purification and antibody preparation—localization in peripheral nerve, artery and testis. *Diabetes* 1980; 29:438–449.

46. Ludvigson MA, Sorenson RL: Immunohistochemical localization of aldose reductase. II. Rat eye and kidney. *Diabetes* 1980; 29:450–459.

47. Sheys GH, Arnold WJ, Watson JA, et al: Aldose reductase from *Rhodotorula*. *J Biol Chem* 1971; 246:3824–3827.

48. Sheys GH, Doughty CC: The reaction mechanism of aldose reductase from *Rhodotorula*. *Biochim Biophys Acta* 1971; 242:523–531.

49. Goil MM, Harpur RP: Aldose reductase and sorbitol dehydrogenase in the muscle of *Ascaris suum* (Nematoda). *Parasitology* 1978; 77:97–102.

50. Halder AB, Crabbe MJC: Bovine lens aldehyde reductase (aldose reductase). Purification, kinetics and mechanism. *Biochem J* 1984; 219:33–39.

51. Feldman HB, Szczepanik PA, Harne P: Stereospecificity of the hydrogen transfer catalyzed by human placental aldose reductase. *Biochim Biophys Acta* 1977; 480:14–20.

52. Crabbe MJC, Halder AB: Kinetic behavior under defined assay conditions for bovine lens aldose reductase. *Clin Biochem* 1979; 12:281–283.

53. Markus HB, Raducha M, Harris H: Tissue distribution of mammalian aldose reductase and related enzymes. *Biochem Med* 1983; 29:31–45.

54. Srivastava SK, Hair GA, Das B: Activated and unactivated forms of human erythrocyte aldose reductase. *Proc Natl Acad Sci USA* 1985; 82:7222–7226.

55. Das B, Hair GA, Srivastava SK: Activated and unactivated forms of aldose reductase and its role in diabetic complications. *Fed Proc* 1985; 44:1391(A).

56. Das B, Srivastava SK: Activation of aldose reductase from tissues. *Diabetes* 1985; 34:1145–1151.

57. Jedziniak JA, Kinoshita JH: Activators and inhibitors of lens aldose reductase. *Invest Ophthalmol* 1971; 10:357–366.

58. Kinoshita JH, Dvornik D, Kraml M, et al: The effect of an aldose reductase inhibitor on the galactose-exposed rabbit lens. *Biochim Biophys Acta* 1968; 158:472–475.

59. Kinoshita JH, Fukushi S, Kador P, et al: Aldose reductase in diabetic complications of the eye. *Metabolism* 1979; 28(suppl I):462–469.

60. Hutton JC, Williams JF, Schofield PJ, et al: Polyol metabolism in monkey-kidney epithelial-cell cultures. *Eur J Biochem* 1974; 49:347–353.

61. Obazawa H, Merola LO, Kinoshita JH: The effects of xylose on the isolated lens. *Invest Ophthalmol Vis Sci* 1974; 13:204–209.

62. Hutton JC, Schofield PJ, Williams JF, et al: The failure of aldose reductase inhibitor 3,3'-tetramethylene glutaric acid to inhibit in vivo sorbitol accumulation in lens and retina in diabetes. *Biochem Pharmacol* 1974; 23:2991–2998.

63. Gabbay KH, Kinoshita JH: Growth hormone sorbitol, and diabetic capillary disease. *Lancet* 1971; 1:913.

64. Chylack LT Jr, Henriques HF, Cheng H-M, et al: Efficacy of alrestatin, an aldose reductase inhibitor, in human diabetic and nondiabetic lenses. *Ophthalmology* 1979; 86:1579.

65. Varma SD, Mukuni I, Kinoshita JH: Flavinoids as inhibitors of lens aldose reductase. *Science* 1975; 188:1215–1216.

66. Dvornik D, Simard-Duquesne N, Kraml M, et al: Polyol accumulation in galactosemic and diabetic rats: Control by an aldose reductase inhibitor. *Science* 1973; 182:1146–1148.

67. Okuda J, Miwa I, Inagaki K, et al: Inhibition of aldose reductases from rat and bovine lenses by flavinoids. *Biochem Pharmacol* 1982; 31:3807–3822.

68. Segelman AB, Segelman FP, Varma SD, et al: *Cannabis sativa L.* (Marijuana) IX: Lens aldose reductase inhibitory activity of marijuana flavone C-glycosides. *J Pharm Sci* 1977; 66:1358–1359.

69. Beyer-Mears A, Farnsworth PN: Diminished diabetic catactogenesis by quercitin. *Exp Eye Res* 1979; 28:709–716.

70. Peterson MJ, Sarges R, Aldinger CE, et al: CP-45,634: A novel aldose reductase inhibitor that inhibits polyol pathway activity in diabetic and galactosemic rats. *Metabolism* 1979; 28(suppl 1):456–461.

71. Fukushi H, Merola L, Kinoshita J: Altering the course of cataracts in diabetic rats. *Invest Opthalmol Vis Sci* 1980; 19:313–315.

72. Poulson R, Boot-Hanford RP, Heath H: Some effects of aldose reductase inhibition upon the eyes of long-term streptozotocin-diabetic rats. *Curr Eye Res* 1982; 2:351–355.

73. Jacobson MJ, Sharma RJ, Cotlier E, et al: Diabetic complications in lens and nerve and their prevention by sulindac or Sorbinil: Two novel aldose reductase inhibitors. *Invest Ophthalmol Vis Sci* 1983; 24:1426–1429.

74. Richon AB, Maragoudakis ME, Wasvary JS: Isoxazolidine-3,5-diones as lens aldose reductase inhibitors. *J Med Chem* 1982; 25:745–747.

75. Sohda T, Mizuno K, Imamiya E, et al: Studies on antidiabetic agents. 5-Arylthiazolidine-2,4-diones as potent aldose reductase inhibitors. *Chem Pharm Bull* 1982; 30:3601–3616.

76. Schnur RC, Sarges R, Peterson MJ: Spiro oxazolidinedione aldose reductase inhibitors. *J Med Chem* 1982; 25:1451–1454.

77. Pfister JR, Wymann WE, Mahoney JM, et al: Synthesis and aldose reductase inhibitor activity of 7-sulfamoylxanthone-2-carboxylic acids. *J Med Chem* 1980; 23:1264–1267.

78. Kador PF, Sharpless NE: Structure-activity studies of aldose reductase inhibitors containing the 4-oxo-4H-chromen ring system. *Biophys Chem* 1978; 8:81–85.

79. Simard-Duquesne N, Greslin E, Dubuc J, et al: The effects of a new aldose reductase inhibitor (tolrestat) in galactosemic and diabetic rats. *Metabolism* 1985; 34:885–892.

80. Sestanj K, Bellini F, Fung S, et al: *N*([5-(trifluoromethyl)-6-methoxy-1-naphthalenyl]thioxomethyl)-*N*-methylglycine (tolrestat), a potent orally active aldose reductase inhibitor. *J Med Chem* 1984; 27:255–256.

81. Inagaki K, Miwa I, Yashiro T, et al: Inhibition of aldose reductases from rat and bovine lenses by hydantoin derivatives. *Chem Pharmacol Bull* 1984; 30:3244–3254.

82. Chaudhry PS, Cabrera J, Juliani HR, et al: Inhibition of human lens aldose reductase by flavinoids, sulindac and indomethacin. *Biochem Pharmacol* 1983; 32:1995–1998.

83. Crabbe MJC, Freeman G, Halder AB, et al: The inhibition of bovine lens aldose reductase by Clinoril, its absorption into the human red cell and its effect on human red cell aldose reductase activity. *Ophthalmic Res* 1985; 17:85–89.

84. Kikkawa R, Hatanaka I, Yasuda H, et al: Effect of a new aldose reductase inhibitor, (E)-3-carboxymethyl-5[(2E)-methyl-3-phenylpropenylidene] rhodanine (ONO-2235), on peripheral nerve disorders in streptozotocin-diabetic rats. *Diabetologia* 1984; 24:290–292.

85. Sharma YR, Cotlier E: Inhibition of lens and cataract aldose reductase by protein bound anti-rheumatic drugs: Salicylate, indomethacin, oxyphenbutazone, sulindac. *Exp Eye Res* 1982; 35:21–27.

86. Kador PF, Sharpless NE, Goosey JD: Aldose reductase inhibition by anti-allergy compounds, in Weiner H, Wermuth B (eds): *Enzymology of Carbonyl Metabolism: Aldehyde Dehydrogenase and Aldo/Keto Reductase*, pp 243–259. New York, Alan R. Liss, 1982.

87. Ono H, Hayano S: 2,2',4'4'-Tetrahydroxybenzophenone as a new aldose reductase inhibitor. *Nippon Ganka Gakkai Zasshi* 1982; 86:353–357.

88. Kador PF, Goosey JD, Sharpless NE, et al: Stereospecific inhibition of aldose reductase. *Eur J Med Chem* 1981; 16:293–298.

89. Varma SD, Kinoshita JH: Inhibition of lens aldose reductase by flavinoids. *Biochem Pharmacol* 1976; 25:2505–2513.

90. Kador PF, Sharpless NE: Pharmacophor requirements of the aldose reductase inhibitor site. *Mol Pharmacol* 1983; 24:521–531.

91. Halder AB, Crabbe MJCC: Inhibition of aldose reductase by phenylglyoxal, diethylpyrocarbonate and thiol modifiers. *Biochem Soc Trans* 1982; 10:401–403.

92. Wolff SP, Crabbe MJC, Thornally PJ: Autoxidation of glyceraldehyde and other simple monosaccharides. *Experientia* 1984; 40:244–246.

93. Thornally PJ, Wolff SP, Crabbe MJC, et al: Autoxidation of glyceraldehyde and other monosaccharides catalyzed by buffer ions. *Biochim Biophys Acta* 1984; 797:276–287.

94. O'Brien MM, Schofield PJ, Edwards MR: Inhibition of human brain aldose reductase and hexonate dehydrogenase by alrestatin and Sorbinil. *J Neurochem* 1982; 39:810–814.

95. Srivastava SK, Petrash JM, Sadana IJ, et al: Susceptibility of aldehyde and aldose reductase of human tissues to aldose reductase inhibitors. *Curr Eye Res* 1982; 2:407–410.

96. Kador PL, Merola LO, Kinoshita JH: Differences in the susceptibility of aldose reductase to inhibition. *Doc Ophthalmol Proc Ser* 1979; 18:117–124.

97. Kador PF, Kinoshita JH, Tung WH, et al: Differences in the susceptibility of aldose reductase to inhibition. *Invest Ophthalmol Vis Sci* 1980; 19:980–982.

98. Yoshida H: The characteristics of aldose reductase in human lens, placenta, and rat organs. *Nippon Ganka Gakkai Zasshi* 1981; 85:865–869.

99. Maragoudakis ME, Wasvary J, Gaigiulo P, et al: Human placental aldose reductase: Sensitive and insensitive forms to inhibition by alrestatin. *Fed Proc* 1979; 30:255(A).

CHAPTER 3

Aldose Reductase and Complications of the Eye

Cataracts

The paradigm for the participation of the polyol pathway in the pathogenesis of a complication of diabetes is the ocular lens, the first tissue in which an association between excess sorbitol formation and a pathologic change was described. The discovery by Von Heyningen that the lenses of rats in which cataracts had been induced by diabetes, or by galactose or xylose feeding, contained increased amounts of the respective sugar alcohols sorbitol, galactitol, and xylitol provided an explanation for certain histopathologic features that had been observed in developing cataracts.[1,2] These features consisted of the early appearance of hydropic lens fibers, followed by rupture of the swollen fibers, liquefaction, and replacement by vacuoles or interfibrillar clefts.

Although the identified sugar alcohols are not themselves toxic to the cell, they are osmotically active substances which can cause transcellular movement of water. It was quickly appreciated that accumulation of sugar alcohols in lens fiber cells could lead to hypertonicity with entry of excess water and cellular overhydration, exaggerated by the fact that these substances penetrate cell membranes poorly and hence, once formed, remain trapped within the cell.[3,4] Indeed, Kinoshita and colleagues demonstrated in a number of studies that polyol accumulation in the lens is accompanied by a parallel increase in water content and lens hydration,[5-12] thus suggesting a mechanism for the initiation of cataract formation that had definite histopathologic correlates, as noted above. The disruption of fiber structure that follows osmotic swelling ultimately leads to loss of transparency.[13] This sequence of polyol accumulation and lenticular swelling also helped explain the long-recognized occurrence of semiacute changes in the refractive index in diabetic patients, with myopia due to lens swelling during hyperglycemic phases and hypermetropia when the blood glucose is lowered.[14-16]

Polyol-induced hypertonicity of the lens promotes entry of sodium ions along with water. As osmotic swelling proceeds, membrane permeability is altered, resulting in loss of potassium ions and amino acid concentrating ability, and efflux of *myo*-inositol.[10,12,17-25] Other changes include decreased ATP levels, loss of reduced glutathione, and leakage of peptides (Figure 3-1). The synthesis of lens crystallins is depressed during development of galactose-induced cataracts, but synthesis of noncrystallin proteins is unaffected; leakage of crystallin proteins from cataractous lenses also occurs.[26] The onset of cataracts in galactosemic and diabetic rats is accompanied by a reduction in NADPH and an increase in $NADP^+$ levels,[27] and progression of lens opacity in galactosemic rats is accompanied by decreased Na/K-ATPase activity.[28] Interestingly, treatment with the aldose reductase inhibitor sorbinil prevents the drop in Na/K-ATPase activity, as has been found in other tissues (see Chapters 4 to 6). In diabetic rats, the lens content of glucose-6-phosphate and glycerol-3-phosphate as well as of sorbitol and fructose is increased.[29-31] Glycerol-3-phosphate, but not glucose-6-phosphate, is normalized by treatment with Sorbinil. The ability of Sorbinil to restore glycerol-3-phosphate prompted the suggestion that aldose reductase inhibition affects a wide network of glycolytic reactions in the lens, possibly by influencing altered redox states of $NAD^+/NADH$ and $NADP^+/NADPH$ that are generated by shifts in the activity of the pentose shunt pathway in the diabetic state.[31]

The decrease in reduced glutathione that is observed during sugar-induced cataract development has been interpreted as evidence

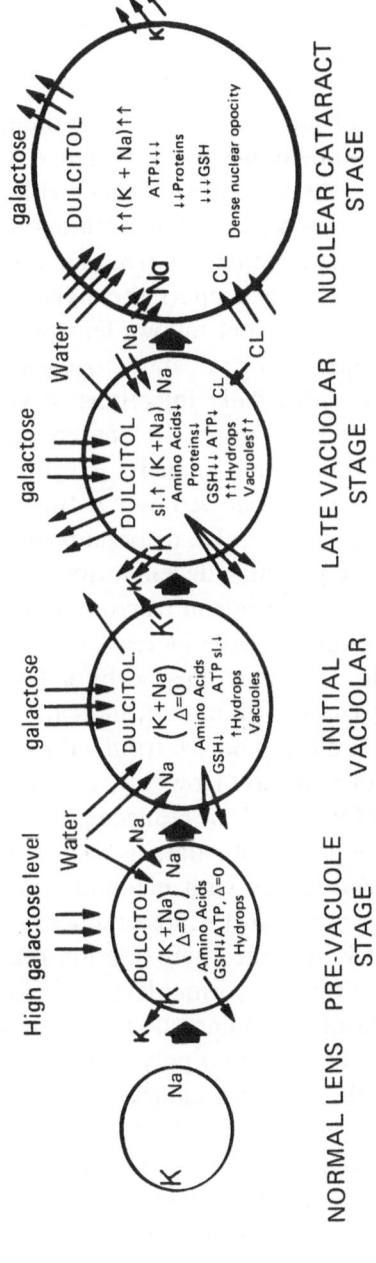

FIGURE 3-1 Sequence of changes involved in the development of sugar cataracts, as proposed by Kinoshita. Reprinted with permission from Kinoshita JH: Cataracts in galactosemia. *Invest Ophthalmol* 1965; 4:786–799.

supporting an alternative or concomitant theory of cataractogenesis, namely, oxygen-dependent oxidative stress on the lens. This theory notes that superoxide and its derivatives can cause lens injury, possibly by polymerization/depolymerization or denaturation reactions involving various macromolecules and by generation of toxic waste products, and that the amount of oxidized proteins is higher in cataractous than in noncataractous lenses.[32-38] The observation that the development of opacities in lens incubated with high glucose concentration is arrested when reduced glutathione or vitamin E, which inhibit superoxide-dependent oxidative reactions, is added to the incubation media seems to support the hypothesis that the decrease in reduced glutathione that follows lens polyol accumulation and swelling is a major factor in cataractogenesis.[39] Similarly, treatment of diabetic rats with daily injections of vitamin E prevented the appearance and evolution of lens changes that were observed in untreated diabetic rats, namely, irregular structure of equatorial fiber cells at 4 days, followed by twisting of and protrusions from these cells at 1 to 2 weeks, disorganization at 3 weeks, and extensive subcapsular globular degeneration of cortical fiber cells at 6 weeks.[40] Notably, the levels of fructose and glucose in the vitamin-treated diabetic animals were increased to the same extent as in untreated diabetic rats, and lens sorbitol levels remained elevated in the rats that received vitamin E, indicating that the protective effect of vitamin E did not derive from a direct influence on the activity of the polyol pathway. However, a more recent study found that the addition of vitamin E or reduced glutathione failed to protect against the development of cataracts that were induced by incubation of rat or gerbil lenses with medium containing a high (55.5 mM) glucose concentration.[41] Aldose reductase inhibitors do not appear to consistently influence peroxidation reactions in diabetic rat lenses,[42] but can restore normal levels of glutathione.[31]

Three additional lines of reasoning, bolstered by experimental evidence, support the hypothesis that polyol accumulation participates in the pathogenesis of sugar cataracts, at least in animal models. First, the development of cataractous changes and the rate of cataract formation, either in vivo in animals with spontaneous or experimental diabetes, or in vitro when lenses are incubated in media containing high glucose concentrations, is greater in lenses in which there is a high level of aldose reductase activity. For example, a mouse model with congenital diabetes and low levels of aldose reductase activity in the lens accumulates little polyol and does not develop cataracts despite persistent hyperglycemia, whereas a rodent model with high levels of aldose reductase activity in the lens develops cataracts within a few weeks after the onset of mild hyperglycemia.[12,43-46] Additionally, lenses from gerbils cultured in medium containing 55.5 mM glucose develop cortical opacities after 24 hours, whereas rat lenses incubated under the same conditions do not develop opacities until 96 hours; aldose reductase activity in gerbil lens is about twice that in rat lens.[41] Gerbils given a high (50%) galactose diet develop cataracts about twice as fast as do galactose-fed rats, and dulcitol accumulation in the lenses of these animals is more pronounced than it is in the lenses of galactosemic rats.[47] Second, cataract formation is accelerated when sugars that have a K_m for aldose reductase that is lower than glucose are used, since they are better substrates, or if the inducing agent is a sugar that forms an alcohol but is not further metabolized to a keto sugar via sorbitol dehydrogenase. Thus, xylose and galactose promote the formation of lenticular opacities or cataracts more efficiently than does glucose.[11,41,43,44,48] Third, prevention of polyol accumulation by inhibition of aldose reductase delays or prevents the onset of cataracts. This has been demonstrated repeatedly in galactose-fed and diabetic animals, and with a variety of compounds, from the earliest available to the most recently developed aldose reductase inhibitors including tetramethyleneglutaric acid[19,49]; Ayerst's alrestatin and tolrestat[50-53]; Pfizer's Sorbinil[54-58]; compounds developed by ICI, Eisai, and ONO[59,60]; flavinoids such as quercitrin[46,61-63]; and the structurally similar xanthones.[64] Although the ability of a variety of other, structurally unrelated compounds such as sulindac (Clinoril®), salicylates, and indomethacin that can act as inhibitors of aldose reductase to forestall the development of sugar cataracts in animal models has not been established, there is interesting clinical information to suggest that they could do so.[65-69]

Treatment with an aldose reductase inhibitor has also been reported to reverse the cataractogenic process after it has been initiated in galactose-fed or streptozotocin-diabetic rats.[70,71] After 5 days of galactose, fiber disintegration and vacuole formation in the pre-equatorial and equatorial cortex were noted; treatment with

Sorbinil despite continued high-galactose feeding during the subsequent 5 days reversed these changes. The lenses of untreated diabetic rats showed increased fiber thickness, edematous granulation of the cell surface, and absence of fiber interdigitation by 16 days after induction of diabetes; subsequent treatment for 5 days with Sorbinil resulted in recovery of fiber contour and interdigitation and appearance of new fibers.

Whole-body irradiation in young rats prolongs the latency period for development of galactose-induced cataracts.[72,73] It is therefore of interest to note the report that whole-body irradiation inhibits the normal age-associated increase in lens aldose reductase activity, either by an effect on enzyme synthesis or by causing production of faulty protein.[74]

A few studies directly support the relevance of findings in experimental sugar cataracts to cataract formation in human diabetes. For example, analysis of cataracts removed from human diabetic subjects has shown that the contents of sorbitol and fructose are increased in specimens removed from patients with elevated fasting blood glucose concentrations[63,75] (Figure 3-2). In one series of patients with diabetes and cataracts, the content of fructose and sorbitol in their cataracts correlated with their fasting blood glucose concentrations and glycosylated hemoglobin levels.[76] Another study, however, found no differences in the lens aldose reductase activity of normal, nondiabetic cataract, and diabetic cataract specimens.[77] Human lenses, obtained from donated eyes kept in an eye bank, synthesize substantial amounts of sorbitol and fructose when incubated in the presence of 30 mM glucose.[51] The tacit assumption that the lens is exposed in vivo to increased concentrations of glucose in human diabetic patients appears to be correct, since the mean glucose concentration in the aqueous humor of cataract patients with diabetes is higher than that in the aqueous humor of nondiabetic subjects.[78] Additionally, increased aqueous humor glucose concentration is reported to elevate the sodium concentration in human lens. However, despite the presence of polyol pathway enzymes in the ciliary body, sorbitol does not accumulate in this tissue in diabetes, at least in rabbits, and enhanced polyol activity thus does not appear to contribute to other diabetes-associated changes in the aqueous humor, such as the reduction in amino acid levels.[79-81]

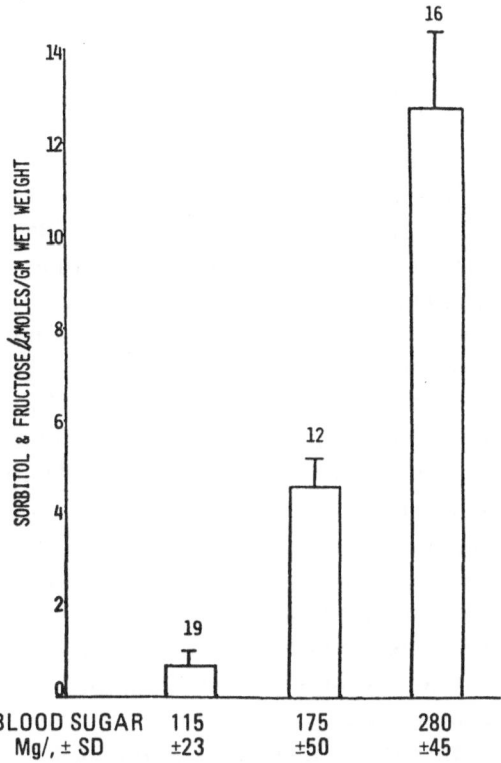

FIGURE 3-2   Relationship between sorbitol plus fructose contents and fasting blood glucose concentrations in cataracts from diabetic subjects. Reprinted with permission from Varma SD et al: Implications of aldose reductase in cataracts in human diabetics. *Invest Ophthalmol Vis Sci* 1979; 18:237–241.

Despite the above considerations, the exact contribution of the polyol pathway to the development of cataracts, either diabetic or senile, in human patients is not entirely clear, and this information must be interpreted in the context of other theories regarding cataractogenesis in diabetic or nondiabetic populations. One of the most prominent among these is that linking the nonenzymatic glycosylation of proteins to cataract formation.[82] That treatment with an aldose reductase inhibitor reportedly prevented cataracts in galactosemic rats without altering the level of nonenzymatic glycosylation of lens proteins could be taken as evidence favoring the primacy of the polyol pathway as a pathogenetic influence.[83] Similarly, Sorbinil has no effect on the level of glucose-6-phosphate, an efficient reactant for nonenzymatic glycosylation, in the lens of diabetic rats.[31] However, these results do not address the issue of animal versus human cataracts, or of the suitability of extrapolating effects induced by galactose, which is not further metabolized to fructose, to those of glucose, which is. These considerations assume particular importance in view of the findings that, on a per lens basis, the level of aldose reductase in human lenses is roughly one order of magnitude lower than that found in many animals, that the specific activity of human lens aldose reductase is less than that of rat or rabbit, and that the ratio of activity of aldose reductase to sorbitol dehydrogenase in human lens is the reverse of that in all other animal lenses that have been studied.[77] Relative to animal lenses, the human lens contains high levels of sorbitol dehydrogenase and limited levels of aldose reductase. Thus, sorbitol is the predominant polyol product in animal lens, but fructose is the main metabolite in human lens. However, fructose could exert a significant osmotic effect, assuming that it is not freely diffusable from the cell and that it is not further metabolized to a great extent. Reports that the prevalence of cataracts in patients receiving aspirin is significantly lower than that in matched populations not receiving aspirin could be taken as evidence that inhibition of nonenzymatic glycosylation (via acetylation of free ϵ-amino groups of lysine residues in lens proteins) prevents cataract formation, but these observations could relate to the aldose reductase-inhibiting properties of salicylates.[68,69,84]

The relationship between sugar-induced cataracts according to the osmotic theory, and cataract formation according to the depletion of reduced glutathione/oxidative dependent oxidative stress theory is not entirely clear, as discussed earlier in this chapter. Further, the potential involvement in cataract formation of abnormalities in the glycolytic pathway or in the metabolic regulation of glycolysis, and how these relate to enhanced polyol pathway activity, have not been fully defined.[85-89] One unifying hypothesis that takes into account the diverse factors that have cataractogenic potential is that which postulates that a combination of two or more such factors may actually be required for development of cataracts. Each factor may be subcataractogenic if operative alone, and may act in concert with or potentiate another, possibly in stepwise fashion.[40,89,90] The contributing factors may not be identical in individual patients. Thus, the decreased lenticular glycolysis, phosphofructokinase, and hexokinase activity, and the excess nonenzymatic glycosylation of lens crystallins that occur with aging may, in combination, contribute to the formation of senile cataracts.[77,85-87] These factors, perhaps coupled with the added stress of sorbitol and fructose accumulation, may explain the increased risk for senile cataracts conferred by the diabetic state.[91] Similarly, it is likely that other and various combinations of stress factors for cataractogenesis pertain in diabetic populations. If this hypothesis is correct, the potential benefit of eliminating at least one subcataractogenic influence, such as polyol accumulation via inhibition of aldose reductase, can be readily appreciated. In fact, a recent study presenting evidence that the addition of $0.1\,mM\ H_2O_2$ to lenses incubated in the presence of $36\,mM$ glucose reduces sorbitol production and accumulation suggests that inhibition of aldose reductase renders the lens better able to cope with oxidative stress.[92] The theory underlying this postulate recognizes that aldose reductase and glutathione reductase both require NADPH as a cofactor and that both oxidation and sorbitol production will activate the hexose monophosphate shunt, the principal source of NADPH in the lens. Thus, activation of the polyol pathway may result in a competition for available NADPH that impairs the tissue's ability to scavenge oxidants.

Retinopathy

The capillaries of the retina contain two types of cells: the endothelial cells and the intramural pericytes.[93-95] Endothclial cells line the capillary lumen and form a permeability (blood-retinal) barrier, derived to a large extent from the tight junctions by which these cells are joined.[96] They produce and are surrounded by a basement membrane, which is continuous with that which envelops the pericyte. Intramural pericytes, also known as mural cells, are selectively lost early in the course of diabetic retinopathy,[97-99] leaving a ghostlike pouch of surrounding basement membrane.

This specific mural cell loss is believed to be a primary event contributory to the retinopathic process, even though it might be obscured by capillary closure later in the course of retinopathy.[100] Breakdown of the blood-retinal barrier is another early feature.[101-103] Dilatation and endothelial cell proliferation occur in the capillaries in which there is a loss of mural cells, while adjacent capillaries have diminished perfusion and eventually become acellular tubes consisting solely of basement membrane.[104] Dilated capillaries develop microaneurysmal outpouchings, leak intravascular fluid, and may eventually give rise to growth and proliferation of new capillaries (neovascularization).

A link between the polyol pathway and the mural cell dropout characteristic of early diabetic retinopathy was first suggested with the demonstration that aldose reductase is present in the retinal microvasculature,[105] and specifically in the mural cells. Immuno-histochemical studies using antibody raised in goats or rabbits against human placental aldose reductase have localized the enzyme in the perinuclear cytoplasm of mural cells of human retinal vessels.[106] In these studies, the successful demonstration of immuno-reactivity was ascribed to the use of short-term trypsin digestion to prepare well-preserved human retinal vessels with intact basement membranes, since two previous studies of frozen retinal cross sections had failed to find or had shown only scattered histochemical evidence of the enzyme in rat and dog retinal vessels.[107,108] Another, more recent study also failed to find aldose reductase immunohistochemically in frozen sections of the retinal vessels, but it was impossible to identify specific mural cell bodies in these sections of retinal capillaries.[109] Aldose reductase activity has also been demonstrated by radioimmunoassay in rhesus monkey mural cells that have been maintained in tissue culture.[110] The radio-immunoassay utilized rabbit antibody that was prepared against purified human placental aldose reductase, and cross-reacted with the enzyme in rhesus lens. It was able to detect aldose reductase in whole retinas as well as in cultured mural cells, but not in freshly isolated retinal capillaries, perhaps because of the paucity of peri-cytes in such preparations and the consequent low relative concen-tration of aldose reductase (Table 3-1).

When mural cells from rhesus monkey are maintained in tissue culture medium that contains a high (40 mM) glucose concentra-tion, sorbitol accumulation occurs. Retinal mural cells also contain higher levels of fructose when grown in 40 mM glucose than they do

TABLE 3-1   Aldose Reductase in Retinal Tissue From
Rhesus Monkeys

Sample	Aldose Reductase (ng/mg protein)
Retinal capillaries	Negligible
Whole retina	69
Cultured mural cells	441

Data from Buzney et al.[110]

when the culture medium contains 10 mM glucose, indicating that
these cells contain sorbitol dehydrogenase activity. The sorbitol
accumulation is evident after 3 days of culture in 40 mM glucose,
rises dramatically between the third and the seventh day of exposure
to high glucose, and thereafter plateaus (Figure 3-3). Retinal

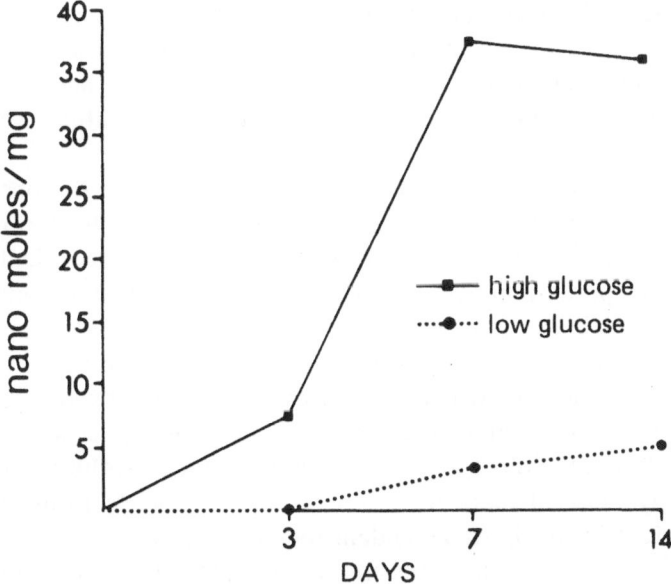

FIGURE 3-3   Sorbitol accumulation in cultured retinal mural cells. High glucose,
40 mM concentration; low glucose, 10 mM concentration. Reprinted with permission from Buzney SM et al: Aldose reductase in retinal mural cells. *Invest
Ophthalmol Vis Sci* 1977; 16:392–396.

TABLE 3-2   Galactitol Production by Retinal Microvessels

Condition	Galactitol μmol/g protein)
30 mM galactose	24.5 $\pm$ 4.1

30 mM galactose plus
   2.5 $\times$ 10^{-5}
   Sorbinil
	11.0 $\pm$ 1.3

Data from Kern and Engerman.[111]

microvessels isolated from dogs and incubated with a high (30 mM) galactose concentration produce dulcitol (galactitol), the hexitol resulting from aldose reductase activity on galactose substrate.[111] This galactitol production is prevented when an aldose reductase inhibitor is included in the incubation medium (Table 3-2).

Aldose reductase activity has been demonstrated by enzymatic assay and by immunologic techniques in a retinoblastoma cell line maintained in tissue culture.[112] Unfortunately, the exact retinal cell of origin of this cell line is not known, making the findings difficult to interpret from the perspective of pericyte dropout and diabetic retinopathy. The retinoblastoma cells accumulate dulcitol when placed in media containing 30 mM galactose, and this accumulation is prevented by inclusion of the aldose reductase inhibitor Sorbinil in the tissue culture media.

The foregoing observations have helped formulate the postulate that polyol accumulation occurs in, and is injurious to, the retinal microvascular mural cell in diabetes. It appears that retinal peri-cytes, like peripheral nerves and other tissues affected by complications of diabetes, do not require insulin for glucose transport. Cultured bovine retinal capillary pericytes have been found to possess a saturable, stereospecific, carrier-facilitated transport mechanism for D-glucose and its analog, 3-O-methyl-D-glucose.[113] These recent findings help support the fundamental hypothesis that the hyperglycemia of diabetes, via shunting of glucose from insulin-dependent to insulin-independent metabolic processes such as the polyol pathway, damages the mural cell. Indeed, polyol content is increased in the retinas of animals with experimental diabetes[114,115] (Figure 3-4). Further, mural cells maintained in tissue culture with medium containing a high glucose concentration have been noted to undergo degenerative changes,[110] consisting of clumping and multilayering of the cells amid extracellular debris, whereas cells grown in 10 mM glucose remained in monolayer and the cultures contained minimal cellular debris (Figure 3-5). The cell multiplication rate and the mitotic rate are reduced when bovine retinal microvessel pericytes are grown in 20 mM compared to 5 mM glucose, but high glucose concentration stimulates protein and collagen synthesis by these cells.[116]

Three additional findings in retinal tissue help support the hypothesis that polyol accumulation contributes to the retinopathic process. First, dogs maintained on a high-galactose diet sufficient to produce galactosemia develop retinal microaneurysms reminiscent of those found in typical diabetic retinopathy[117] (Figure 3-6). Second, the basement membrane in retinal capillaries of rats fed a high-galactose diet for 28 to 44 weeks is thickened, and this thickening is prevented by including an aldose reductase inhibitor in the diet[118] (Figure 3-7). However, these latter findings have to be reconciled with the report that the increased rate of collagen synthesis induced by culture of bovine retinal capillary pericytes in high glucose concentration is not corrected by inclusion in the incubation media of the aldose reductase inhibitor Sorbinil[119] (Figure 3-8). Third, fructose-fed diabetic rats develop retinal microvascular changes that include pericyte loss, microaneurysms, endothelial proliferation, and capillary basement membrane thickening, and treatment with the aldose reductase inhibitor ONO-2235 reduces the severity of these changes.[120]

It is not clear whether polyol accumulation per se, or associated metabolic changes, is (are) deleterious to the retinal microvasculature, and if so, how. According to one group of investigators, the absolute level of sorbitol that accumulates in the retinas of diabetic rats (1.5 μmol/g protein) is insufficient to be of osmotic significance.[114] However, another group noted that, since aldose reductase activity is sequestered in specific cellular locations, accumulated sorbitol might not be evenly distributed throughout the retina and could be a significant osmotic force in individual cells.[107] Further, retinal sorbitol concentrations may have been underestimated in some studies, since levels are considerably higher (68.5 and 243 μmol/100 g tissue; normal and diabetic sucrose-fed rats, respectively) in specimens dissected in situ and placed in perchloric acid within 5 seconds.[121] On the other hand, it has been argued that

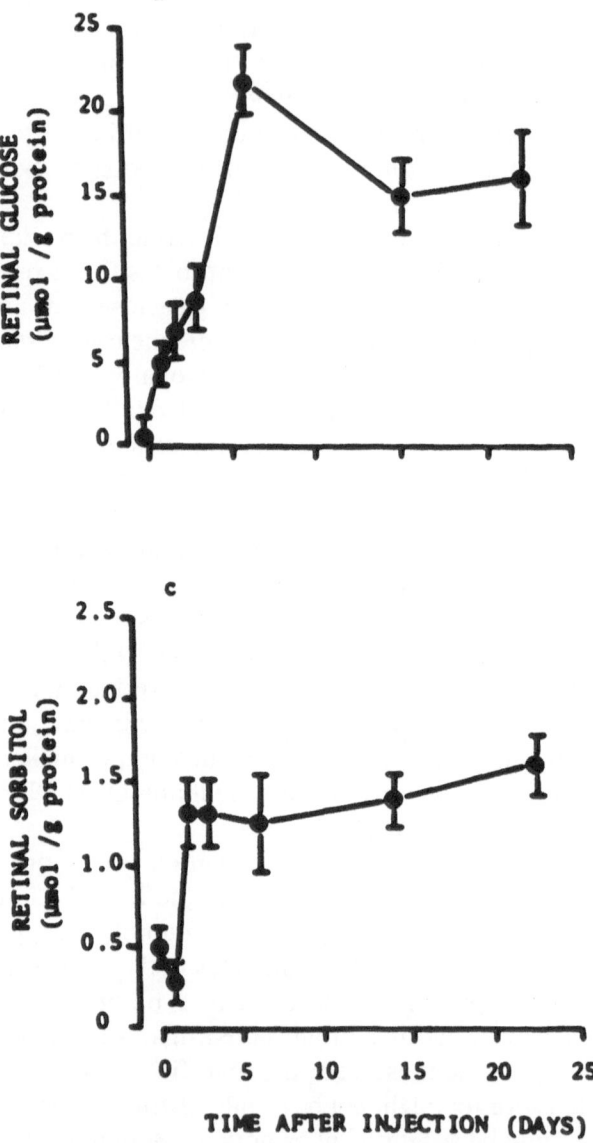

FIGURE 3-4   Concentrations of free carbohydrates in the retina of rats with streptozotocin diabetes. Reprinted with permission from Hutton JC et al: Sorbitol
metabolism in the retina: Accumulation of pathway intermediates in streptozotocin
induced diabetes in the rat. *Aust J Exp Biol Med Sci* 1974; 52:361–373.

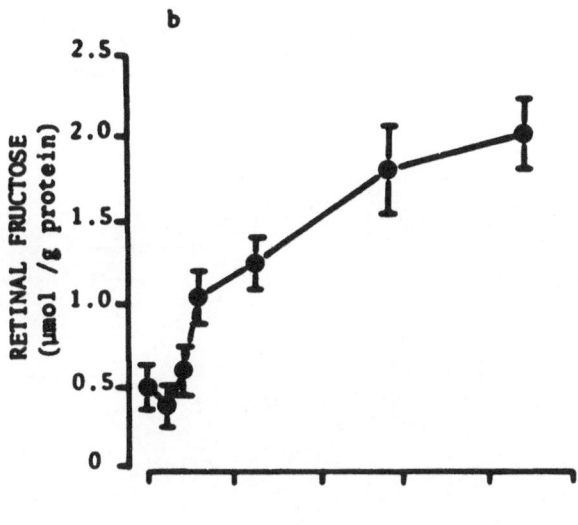

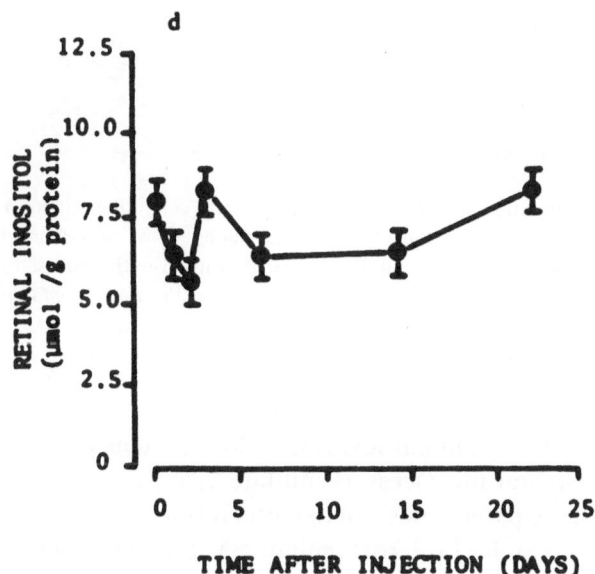

TIME AFTER INJECTION (DAYS)

FIGURE 3-4   *Continued.*

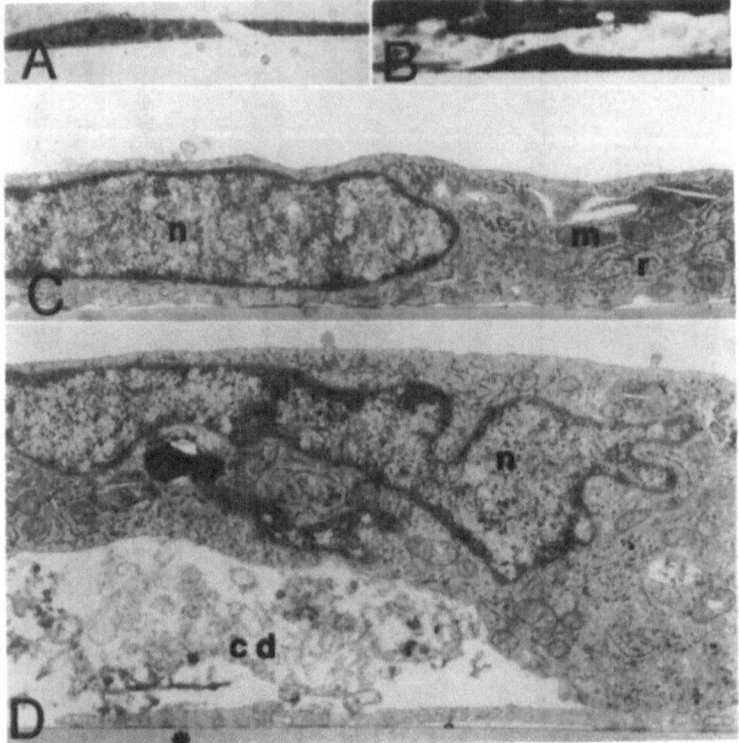

FIGURE 3-5   Electron micrograph of retinal mural cells maintained in 10 mM (C) and 40 mM (D) glucose concentration; n, nucleus; m, mitochondria; r, rough endoplasmic reticulum; cd, cellular debris. Reprinted with permission from Buzney SM et al: Aldose reductase in retinal mural cells. *Invest Ophthalmol Vis Sci* 1977; 16:392–396.

the quantitatively similar levels of aldose reductase activity in retinal and cerebral microvessels militate against a major role of this enzyme in the pathogenesis of diabetic retinopathy, since the retinal but not the cerebral microvasculature is typically damaged in diabetes.[122]

The possibility that other changes in cell nutrients or enzymatic activities accompanying polyol accumulation in the diabetic retina contribute to the pathogenesis of retinopathy is intriguing. One group of investigators found that increased retinal sorbitol in alloxan-

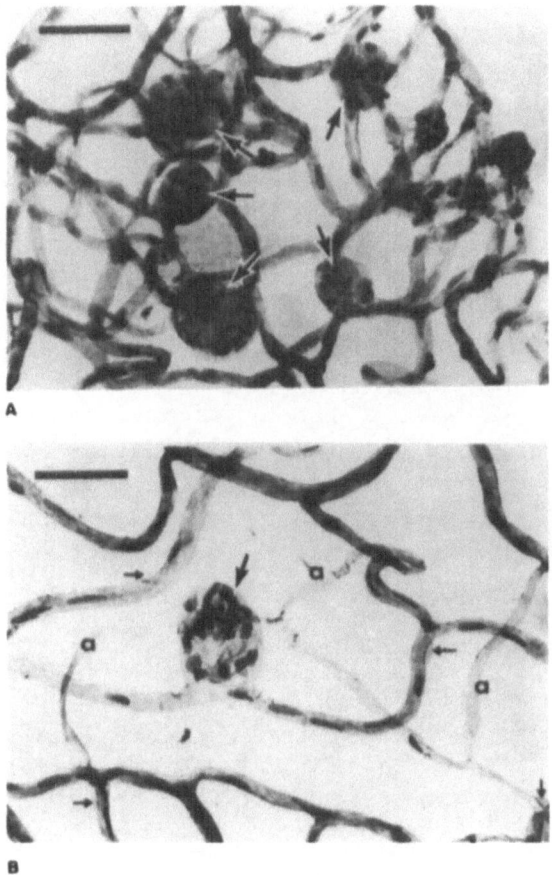

FIGURE 3-6 Retinal vascular changes in experimental galactosemia. A, Capillary aneurysms (large arrows); B, acellular capillaries (a) and ghosts of intramural pericytes (small arrows). Reproduced with permission from the American Diabetes Association, Inc. From Engerman RL, Kern TS: Experimental galactosemia produces diabetic-like retinopathy. *Diabetes* 1984; 33:97–100.

diabetic rabbits is associated with *myo*-inositol depletion and reduced Na/K-ATPase activity,[115] as has been observed in other tissues, notably peripheral nerve and renal glomeruli. Another group, however, noted that rat retinal *myo*-inositol concentrations were unaffected by streptozotocin diabetes.[114] Interestingly, the polyol accumulation induced in cultured retinal mural cells by exposing

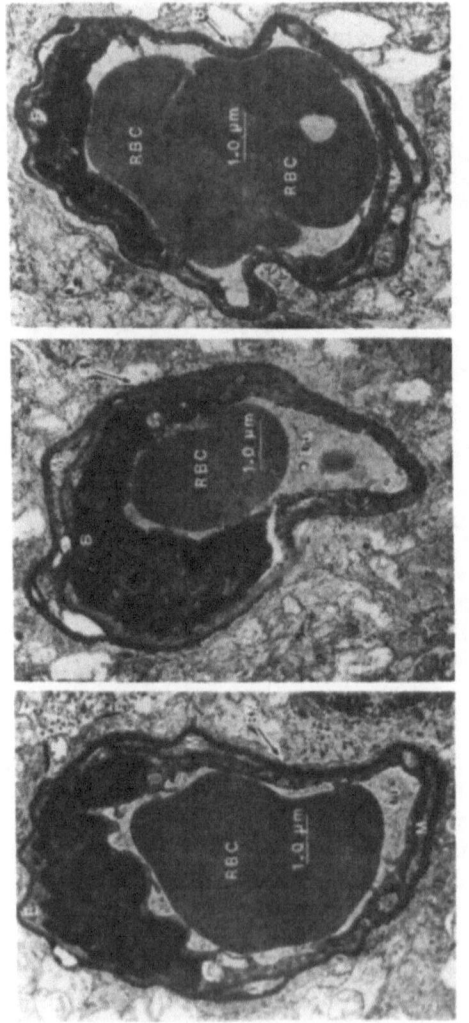

FIGURE 3-7   Electron micrographs of retinal capillaries from control (left), galactose-fed (middle), and Sorbinil-treated, galactose-fed (right) rats. BM, Basement membrane; E, endothelial cell; Lu, lumen; M, mural cell; RBC, red blood cell. Copyright 1983 by the AAAS. From Robison WG et al: Retinal capillaries: Basement membrane thickening by galactosemia prevented with aldose reductase inhibitor. *Science* 1983; 221:1177–1179.

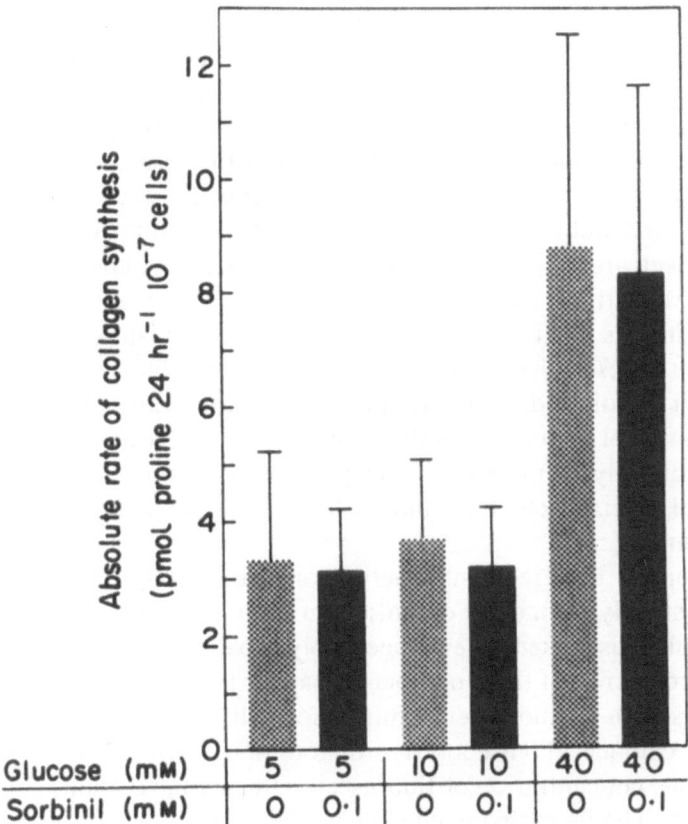

FIGURE 3-8 Collagen synthesis in cultured bovine retinal capillary pericytes. Shaded bars, without Sorbinil; Solid bars, with Sorbinil. Reprinted with permission from Li W et al: The effects of glucose and an aldose reductase inhibitor on the sorbitol content and collagen synthesis of bovine retinal capillary pericytes in culture. *Exp Eye Res* 1985; 40:439–444.

them to high glucose concentration is not accompanied by reduced inositol levels.[110] Nevertheless, the reported changes in retinal *myo*-inositol and Na/K-ATPase activity in diabetes were prevented with inhibition of aldose reductase, suggesting that they are in some way consequent to the enhanced polyol pathway activity. Further, they are associated with a functional alteration, manifest as a reduction in the electroretinogram C wave, that also corrects with aldose reductase inhibition or with *myo*-inositol supplementation.[123,124]

Although the reduced oscillatory potential of the electroretinogram observed after 12 to 16 weeks of diabetes in the rat was not corrected by an aldose reductase inhibitor, the drug significantly improved the change in peak latency shortening of the electroretinogram observed after 8 weeks of diabetes.[125] The relationships between mural cell dropout and/or endothelial cell proliferation and the abnormalities in *myo*-inositol content, Na/K-ATPase activity, and deterioration of the electroretinogram need to be delineated. Studies examining these questions will undoubtedly proceed in conjunction with experiments seeking to unravel analogous relationships that exist in other tissues subject to diabetic complications. The recent description of a sodium-dependent, glucose- inhibited transport system for *myo*-inositol in retinal capillary pericytes[126] suggests that mechanisms underlying metabolic, and perhaps functional, defects in the retinal microvasculature and in the peripheral nerve in diabetes are similar.

There is disagreement whether one abnormal retinal function consistently found in diabetic patients[102,127] and experimental animals[103] is linked to enhanced polyol pathway activity, namely, the breakdown of the blood-ocular barrier to the passage of organic anions such as fluorescein. This abnormality can be quantitated by the technique of vitreous fluorophotometry, which measures the vitreous concentration of fluorescein after its intravenous administration. The increased ocular accumulation of fluorescein observed in rats with streptozotocin diabetes was not duplicated by rendering rats galactosemic with a diet containing 50% galactose, and treatment of diabetic rats with either of two different aldose reductase inhibitors failed to correct the abnormal vitreous fluorophotometric measurements associated with untreated experimental diabetes of 3 weeks duration.[128] On the other hand, treatment for 6 months with sulindac (Clinoril) or Sorbinil, two inhibitors of aldose reductase, was reported to significantly lower leakage on vitreous fluorophotometry in patients with insulin-dependent and non-insulin-dependent diabetes.[129,130] However, the relationship between breakdown of the blood-retinal barrier, as assessed by vitreous fluorophotometry, and the traditional microaneurysmal and neovascular lesions of diabetic retinopathy is not clear. In fact, the former may reflect a structural defect in the retinal pigment epithelium and/or a decreased transport of fluorescein out of the eye, and may not be part of the same putative process that promotes capillary leakage in the retinal microvasculature or in other sites of microangiopathic complications. Nevertheless, the report that Sorbinil diminished the rate of progression over a 6-month period of the penetration of fluorescein across the blood-retinal barrier in 32 patients with non-insulin-dependent diabetes has encouraged the suggestion that aldose reductase inhibition may alter the course of early diabetic retinopathy.[131]

The retina contains sorbitol dehydrogenase as well as aldose reductase, and hence generates increased amounts of fructose when the polyol pathway is activated.[110] It has been proposed that one way in which enhanced retinal polyol metabolism might contribute to the retinopathic process is by rapid and uncontrolled channeling of excess fructose, after phosphorylation, into the glycolytic pathway. This would raise the concentration of retinal lactate, believed by some to be of significance in the development of diabetic retinopathy.[132,133] In this context, it is interesting to note that one aldose reductase inhibitor can reduce lactate output in cultured kidney epithelial cells incubated for several hours in a high (55 mM) glucose concentration,[134] and that another, given in vivo, can normalize the concentration of lactate in the retinas of streptozotocin- diabetic rats.[135] However, this compound, ICI 10552, had no significant effect on the elevated retinal concentrations of sorbitol and fructose in diabetes.

Keratopathy

Although a variety of corneal lesions are known to occur spontaneously in diabetes,[136,137] interest in diabetic keratopathy, as these abnormalities are collectively called, was spurred with the advent of vitrectomy for the treatment of certain retinal complications of diabetes. Since the cornea in diabetic subjects heals slowly even with minor trauma,[138,139] and since the corneal epithelium often is intentionally removed during vitrectomy, the problem soon became apparent. Several reports have described the propensity for the development of corneal abnomalities in diabetic patients after vitreous surgery.[140-144] These are ascribed to a delay in the re-epithelialization of the cornea after surgical manipulation, resulting in a persistent corneal epithelial defect and, occasionally, stromal edema.

The possibility that increased polyol pathway activity might contribute to diabetic keratopathy was dramatically highlighted in a recent case report that describes the response of severe keratopathy in a young woman with insulin-dependent diabetes to treatment with an aldose reductase inhibitor.[145] The woman had bilateral persistent corneal epithelial defects complicated by reduced corneal sensation and keratitis sicca. Response to conventional therapy for more than a year was inadequate, with recurrence of the corneal defects and intraocular inflammation. After the patient had received Sorbinil in one eye and placebo in the other for 2 months, only the Sorbinil-treated eye showed diminution in the size of the epithelial defects and in the signs of inflammation. Subsequently, both eyes received Sorbinil, resulting in marked clinical improvement.

Aldose reductase has been localized immunohistochemically, using specific antibodies raised in rabbit and goat against purified human placental aldose reductase, in the human corneal endothelium and epithelium.[146] The presence of the sorbitol pathway and its activation in vitro by high glucose concentration have been demonstrated in the rabbit corneal epithelium.[147] In this tissue, glucose uptake and glycogen formation are the same without or with insulin added in vitro, indicating that corneal epithelial cells are insulin-insensitive and thereby suggesting that they would accumulate sorbitol if exposed to a hyperglycemic milieu. Bovine corneal epithelium incubated in xylose-containing medium accumulates xylitol, and inclusion of an aldose reductase inhibitor in the

TABLE 3-3   Sugar Contents in Human Corneal
Epithelium

Sugar	uM/g dry wt	
	Normal	Diabetic
Fructose	—	2.2
Glucose	1.8	12.2
Sorbitol	—	0.62

Data from Foulkes et al.[142]

FIGURE 3-9 Re-epithelialization of cornea after denudation in normal (A), galactosemic (B), and Sorbinil-treated galactosemic (C) rats. Reprinted with permission from Datiles MD et al: Corneal re-epithelialization in galactosemic rats. *Invest Ophthalmol Vis Sci* 1983; 24:563–569.

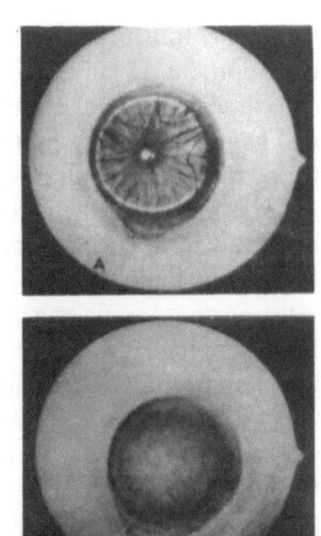

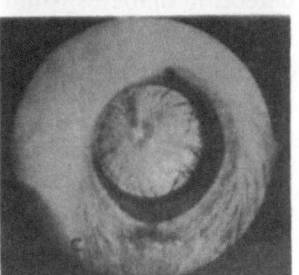

incubation prevents the synthesis of xylitol.[12] Corneal epithelium from diabetic patients reportedly contains elevated levels of glucose, sorbitol, and fructose, as well as two other unidentified sugars that are not present in the epithelium of nondiabetic patients[136,142] (Table 3-3). Thus, aldose reductase functions in the corneal epithelium and can be activated by increased intracellular concentrations of sugar. It has been proposed that sorbitol accumulation in corneal epithelial cells results in osmotic changes that make the diabetic tissue more vulnerable than usual to intraoperative damage,[138] although it is arguable whether the sorbitol level is sufficient to produce an osmotic effect.[147]

Support for the hypothesis that aldose reductase is involved in the development of keratopathic lesions is derived from studies with experimental animal models of diabetes and galactosemia. Diabetic and galactosemic rats both exhibit delayed healing after scraping of the corneal epithelium; treatment with an aldose reductase inhibitor restored the rate of healing of the denuded corneal epithelium to that observed in normal animals.[12,145,148,149] The ability of aldose reductase inhibition to hasten regeneration of the epithelium in diabetic rats was not specific to a particular compound since several different inhibitors had similar beneficial effects. Inhibition of aldose reductase also promoted healing, characterized by a clear and transparent cornea, in contrast to the cloudiness and edema found in the re-epithelialized cornea of untreated diabetic or galactosemic rats[148,149] (Figure 3-9).

The manner in which the polyol pathway participates in the development of diabetic keratopathy or the persistence of corneal defects in diabetes is unknown. Nevertheless, the single description of a dramatic clinical response and the report of a significant experimental response to inhibition of aldose reductase activity in the cornea suggest that the accumulation of polyol products or consequences thereof are involved. These findings offer promise that aldose reductase inhibitors might beneficially influence corneal abnormalities in diabetes, and should encourage further work to assess the therapeutic potential of these agents for diabetic keratopathy.

References

1. Van Heyningen R: Formation of polyols by the lens of the rat with sugar cataracts. *Nature (Lond)* 1959; 184:194–195.

2. Van Heyningen R: Metabolism of xylose by the lens. Rat lens in vivo and in vitro. *Biochem J* 1959; 73:197–207.

3. LeFevre PG, Davis R: Active transport into the human erythrocyte: Evidence from comparative kinetics and composition among monosaccharides. *J Gen Physiol* 1951; 34:515–524.

4. Wick AN, Drury DR: Insulin and permeability of cells to sorbitol. *Am J Physiol* 1951; 166:421–423.

5. Kinoshita JH, Merola LO, Satoh K, et al: Osmotic changes caused by the accumulation of dulcitol in the lenses of rats fed with galactose. *Nature (Lond)* 1962; 194:1085–1087.

6. Kinoshita JH, Merola LO, Dikmak E: Osmotic changes in experimental galactose cataracts. *Exp Eye Res* 1962; 1:405–410.

7. Kinoshita JH, Merola LO, Dikmak E: The accumulation of dulcitol and water in rabbit lens incubated with galactose. *Biochim Biophys Acta* 1962; 62:176–178.

8. Kinoshita JH, Merola LO: Hydration of the lens during the development of galactose cataract. *Invest Ophthalmol* 1964; 3:577–584.

9. Kinoshita JH, Futterman S, Satoh K, et al: Factors affecting the formation of sugar alcohols in ocular lens. *Biochim Biophys Acta* 1963; 74:340–350.

10. Kinoshita JH: Cataracts in galactosemia. *Invest Ophthalmol* 1965; 4:786–799.

11. Kinoshita JH: Mechanisms initating cataract formation. *Invest Ophthalmol* 1974; 13:713–724.

12. Kinoshita JH, Fukushi S, Kador P, et al: Aldose reductase in diabetic complications of the eye. *Metabolism* 1979; 28:462–469.

13. Kuwabara T, Kinoshita JH, Cogan DC: Electron microscopic study of galactose-induced cataract. *Invest Ophthalmol* 1969; 8:133–149.

14. Duke-Elder WS: Changes in refraction in diabetes mellitus. *Br J Ophthalmol* 1925; 9:167–187.

15. Vere DW, Verrel D: Relation between blood sugar level and the optical properties of the lens of the human eye. *Clin Sci* 1955; 14:183–196.

16. Varma SD, El-Aguizy HK, Richards RD: Refractive changes in alloxan diabetic rabbits. Control by flavinoids. *Acta Ophthalmol* 1980; 58:748–759.

17. Kinoshita JH, Merola LO, Hayman S: Osmotic effects on the amino acid-concentrating mechanism in the rabbit lens. *J Biol Chem* 1965; 240:313–315.

18. Kinoshita JH, Merola O, Tung B: Changes in cation permeability in the galactose-exposed rabbit lens. *Exp Eye Res* 1968; 7:80–90.

19. Chylack LT Jr, Kinoshita JH: A biochemical evaluation of a cataract induced in a high glucose medium. *Invest Ophthalmol* 1969: 8:401–412.

20. Varma SD, Kinoshita JH: Sorbitol pathway in diabetic galactosemic rat lens. *Biochim Biophys Acta* 1974; 338:632–640.

21. Cotlier E: *Myo*-inositol: Active transport by the crystalline lens. *Invest Ophthalmol* 1970; 9:681–691.

22. Broekhuyse RM: Changes in *myo*-inositol permeability in the lens due to cataractous conditions. *Biochim Biophys Acta* 1968; 163:269–272.

23. Sippel TO: Changes in the water, protein and glutathione contents of the lens in the course of galactose cataract development in rats. *Invest Ophthalmol* 1960; 5:568–575.

24. Kinoshita JH, Barber GW, Merola LD, et al: Changes in the levels of free amino acids and *myo*-inositol in the galactose-exposed lens. *Invest Ophthalmol* 1969; 8:625–632.

25. Reddy VN, Schauss D, Chakrapani B, et al: Biochemical changes associated with the development and reversal of galactose cataracts. *Exp Eye Res* 1976; 23:483–493.

26. Kador PF, Zigler S, Kinoshita JH: Alterations of lens protein synthesis in galactosemic rats. *Invest Ophthalmol Vis Sci* 1979; 18:696–702.

27. Lee SM, Schade SZ, Doughty CC: Aldose reductase, NADPH and NADP$^+$ in normal, galactose-fed and diabetic rat lens. *Biochim Biophys Acta* 1985; 841:247–253.

28. Unakar NJ, Tsui JY: Inhibition of galactose-induced alterations in ocular lens with Sorbinil. *Exp Eye Res* 1983; 36:685–694.

29. Gonzalez AM, Sochor M, McLean P: Effect of experimental diabetes on glycolytic intermediates and regulation of phosphofructokinase in rat lens. *Biochem Biophys Res Commun* 1980; 95:1173–1179.

30. Gonzalez AM, Sochor M, Rowles PM, et al: Sequential biochemical and structural changes occurring in rat lens during cataract formation in experimental diabetes. *Diabetologia* 1981; 21:5.

31. Gonzalez AM, Sochor M, McLean P: The effect of an aldose reductase inhibitor (Sorbinil) on the level of metabolites in lenses of diabetic rats. *Diabetes* 1983; 32:482–485.

32. Bhuyan KC, Bhuyan DK, Katzin DM: Amizol-induced cataract and inhibition of lens catalase in rabbit. *Ophthalmic Res* 1973; 5:236–247.

33. Reddy VN: Metabolism of glutathione in the lens. *Exp Eye Res* 1971; 11:310–328.

34. Varma SD, Kumar S, Richards RD: Protection by ascorbate against superoxide injury to the lens. *Invest Ophthalmol Vis Sci* 1979; 18(suppl):98.

35. Varma SD: Superoxide and lens of the eye. A new theory of cataractogenesis. *Int J Quantum Chem* 1981; 20:479–484.

36. Goosey JD, Zigler JS Jr, Kinoshita JH: Cross-linking of lens crystallins in a photodynamic system. A singlet oxygen mediated process. *Science* 1980; 208:1278–1279.

37. Dische Z, Zil H: Studies on the oxidation of cysteine to cystine in lens protein during cataract formation. *Am J Ophthalmol* 1951; 34:104–113.

38. Garner MH, Spector A: Selective oxidation of cysteine and methionine in normal and cataractous lens. *Proc Natl Acad Sci USA* 1980; 77:1274–1277.

39. Creighton MO, Trevithick JR: Cortical cataract formation prevented by vitamin E and glutathione. *Exp Eye Res* 1979; 29:689–693.

40. Ross WM, Creighton MO, Stewart-DeHaan PJ, et al: Modelling cortical cataractogenesis: 3. In vivo effects of vitamin E on cataractogenesis in diabetic rats. *Can J Ophthalmol* 1982; 17:61–66.

41. Chand D, El-Aguizy K, Richards RD, et al: Sugar cataracts in vitro: Implications of oxidative stress and aldose reductase I. *Exp Eye Res* 1982; 35:491–497.

42. Kadoya K, Hashi H, Yui MNH, et al: Influences of aldose reductase inhibitor on peroxidation reaction in the lens of streptozotocin diabetic rats. *Nippon Ganka Kiyo* 1983; 34:2172–2176.

43. Cogan DG, Kinoshita KH, Kador PF, et al: Aldose reductase and complications of diabetes. *Ann Intern Med* 1984; 101:82–91.

44. Gabbay KH: The sorbitol pathway and complications of diabetes. *N Engl J Med* 1973; 288:831–836.

45. Varma SD, Kinoshita JH: The absence of cataracts in mice with congenital hyperglycemia. *Exp Eye Res* 1974; 19:577–582.

46. Varma SD, Mikuni I, Kinoshita JH: Flavinoids as inhibitors of lens aldose reductase. *Science* 1975; 188:1215–1216.

47. El-Aguizy HK, Richards RD, Varma SD: Sugar cataracts in mongolian gerbil (*Meriones unguiculatus*). *Exp Eye Res* 1983; 36:839–844.

48. Obazawa H, Merola LO, Kinoshita JH: The effects of xylose on the isolated lens. *Invest Ophthalmol* 1974; 13:204–209.

49. Kinoshita JH, Dvornik D, Kraml M, et al: The effect of an aldose reductase inhibitor on the galactose-exposed rabbit lens. *Biochim Biophys Acta* 1968; 158:472–475.

50. Dvornik D, Simard-Duquesne N, Kraml M, et al: Polyol accumulation in galactosemic and diabetic rats: Control by an aldose reductase inhibitor. *Science* 1973; 182:1146–1148.

51. Chylack LT Jr, Henriques HF, Cheng H-M, et al: Efficacy of alrestatin, an aldose reductase inhibitor, in human diabetic and nondiabetic lenses. *Ophthalmology* 1979; 86:1579–1585.

52. Simard-Duquesne N, Greslin E, Gonzalez R, et al: Prevention of cataract development in severely galactosemic rats by the aldose reductase inhibitor, tolrestat. *Proc Soc Exp Biol Med* 1985; 178:599–605.

53. Simard-Duquesne N, Greslin E, Dubuc J, et al: The effect of a new aldose reductase inhibitor (tolrestat) in galactosemic and diabetic rats. *Metabolism* 1985; 34:885–892.

54. Peterson MJ, Sarges R, Aldinger CE, et al: CP-45,634: A novel aldose reductase inhibitor that inhibits polyol pathway activity in diabetic and galactosemic rats. *Metabolism* 1979; 28(suppl 1):456–461.

55. Beyer-Mears A, Cruz E, Nicolas-Alexandre J, et al: Sorbinil protection of lens protein components and cell hydration during diabetic cataract formation. *Pharmacology* 1982; 24:193–200.

56. Datiles M, Fukui H, Kuwabara T, et al: Galactose cataract prevention with Sorbinil, an aldose reductase inhibitor: A light microscopic study. *Invest Ophthalmol Vis Sci* 1982; 22:174–179.

57. Peterson MJ, Sarges R, Aldinger CE, et al: Inhibition of polyol pathway activity in diabetic and galactosemic rats by the aldose reductase inhibitor CP-45,634. *Adv Exp Med Biol* 1979; 119:347–356.

58. Fukushi H, Merola L, Kinoshita JH: Altering the course of cataracts in diabetic rats. *Invest Ophthalmol Vis Sci* 1980; 19:313–315.

59. Poulsom R, Boot-Hanford RP, Heath H: Some effects of aldose reductase inhibition upon the eyes of long-term streptozotocin-diabetic rats. *Curr Eye Res* 1982; 2:351–355.

60. Ono H, Nozawa Y, Hayano S: Effects of M-79,175, an aldose reductase inhibitor, on experimental sugar cataracts. *Nippon Ganka Gakkai Zasshi* 1982; 86:1343–1350.

61. Beyer-Mears A, Farnsworth PN: Diminished diabetic cataractogenesis by quercetin. *Exp Eye Res* 1979; 28:709–716.

62. Varma SD, Mizuno A, Kinoshita JH: Diabetic cataracts and flavinoids. *Science* 1977; 195:205–206.

63. Varma SD, Shockert SS, Richards RD: Implications of aldose reductase in cataracts in human diabetics. *Invest Ophthalmol Vis Sci* 1979; 18:237–241.

64. Beyer-Mears A, Cruz E, Nicolas-Alexandre J, et al: Xanthone-2-carboxylic acid effect on lens growth, hydration and proteins during diabetic cataract development. *Arch Int Pharmacodyn Ther* 1982; 259:166–176.

65. Sharma YR, Cotlier E: Inhibition of lens and cataract aldose reductase by protein-bound anti-rheumatic drugs: Salicylate, indomethacin, oxyphenbutazone, sulindac. *Exp Eye Res* 1982; 35:21–27.

66. Jacobson M, Sharma YR, Cotlier E, et al: Diabetic complications in lens and nerve and their prevention by sulindac or Sorbinil: Two novel aldose reductase inhibitors. *Invest Ophthalmol Vis Sci* 1983; 24:1426–1429.

67. Crabbe MJC, Freeman G, Halder G, et al: The inhibition of bovine lens aldose reductase by Clinoril, its absorption into human red cell and its effect on human red cell aldose reductase activity. *Ophthalmic Res* 1985; 17:85–89.

68. Cotlier E, Fagadau W, Cicchetti DV: Methods for evaluation of medical therapy of senile and diabetic cataracts. *Trans Ophthalmol Soc UK* 1982; 102:416–422.

69. Cotlier E, Sharma YR, Niven T, et al: Distribution of salicylate in lens and intraocular fluids and its effect on cataract formation. *Am J Med* 1983; 75(6A):83–90.

70. Hu T-S, Datiles M, Kinoshita JH: Reversal of galactose cataract with sorbinil in rats. *Invest Ophthalmol Vis Sci* 1983; 24:640–644.

71. Beyer-Mears A, Cruz E: Reversal of diabetic cataract by sorbinil, an aldose reductase inhibitor. *Diabetes* 1985; 34:15–21.

72. Hockwin O, Bergeder HD, Kaiser L: Über die Galaktosekatarakt junger Ratten nach Ganzkorperröntgenbestrahlung. *Ber Dtsch Ophthalmol Ges* 1967; 68:135–139.

73. Hockwin O, Bergeder HD, Ninnemann U, et al: Untersuchungen zur Latenzzeit der Galaktosekatarakt von Ratten. Einfluss von Röntgenbestrahlung und Diätbeginn bei verschieden alten Tieren. *Graefes Arch Klin Opthalmol* 1974; 189:171–178.

74. Keller H-W, Stinnesbeck TH, Hockwin O, et al: Investigations on the influence of whole body X-irradiation on the activity of rat lens aldose reductase (E.C.1.1.1.21). *Graefes Arch Klin Opthalmol* 1981; 215:181–186.

75. Chylack LT Jr, Henriques H, Tung W: Inhibition of sorbitol production in human lenses by an aldose reductase inhibitor. *Invest Ophthalmol Vis Sci* 1978; 17(ARVO suppl):300.

76. Lerner BC, Varma SD, Richards RD: Polyol pathway metabolites in human cataracts. *Arch Ophthalmol* 1984; 102:917–920.

77. Jedziniak JA, Chylack LT, Chen H–M, et al: The sorbitol pathway in the human lens: Aldose reductase and polyol dehydrogenase. *Invest Ophthalmol Vis Sci* 1981; 20:314–326.

78. Davies PD, Duncan G, Pynsent PB, et al: Aqueous humor glucose concentration in cataract patients and its effect on the lens. *Exp Eye Res* 1984; 39:605–609.

79. Reddy DVN, Kinsey VE: Transport of amino acids into intraocular fluids and lens in diabetic rabbits. *Invest Ophthalmol* 1963; 2:237–242.

80. Reddy DVN: Amino acid transport in the lens in relation to sugar cataracts. *Invest Ophthalmol* 1965; 4:700–706.

81. Reddy VN, Chakrapani B, Steen D: Sorbitol pathway in the ciliary body in relation to accumulation of amino acids in the aqueous humor of alloxan diabetic rabbits. *Invest Ophthalmol* 1971; 100:870–875.

82. Cohen MP: *Diabetes and Protein Glycosylation: Measurement and Biologic Relevance.* New York, Springer-Verlag, 1986.

83. Chiou SH, Chylack LT, Bunn HF, et al: Role of nonenzymatic glycosylation in experimental cataract formation. *Biochem Biophys Res Commun* 1980; 95:894–901.

84. Cotlier E: Aspirin effect on cataract formation in patients with rheumatoid arthritis alone or combined with diabetes. *Int Ophthalmol* 1981; 3:173–179.

85. Bous F, Hockwin O, Ohrloff C, et al: Investigation on phosphofructokinase (EC 2.7.1.11) in bovine lens in dependence on age, topographic distribution and water soluble protein fractions. *Exp Eye Res* 1977; 24:383–389.

86. Chen HM, Chylack LT Jr: Factors affecting the rate of lactate production in rat lens. *Ophthalmic Res* 1977; 9:381–387.

87. Ohrloff C, Zierz S, Hockwin O: Investigations of the enzymes involved in the fructose breakdown in the cattle lens. *Ophthalmic Res* 1982; 14:221–229.

88. Siddiqui MA, Rahman MA: Effect of hyperglycemia on the enzyme activities of lenticular tissue of rats. *Exp Eye Res* 1980; 31:463–469.

89. Chylack LT, Cheng H–M: Sugar metabolism in the crystalline lens. *Surv Ophthalmol* 1978; 23:26–34.

90. Hollows JC, Schofield PJ, Williams JF, et al: The effect of an unsaturated fat-diet on cataract formation in streptozotocin-induced diabetic rats. *Br J Nutr* 1976; 36:161–177.

91. Kahn HA, Liebowitz HM, Ganley JP, et al: The Framingham eye study. II. Association of ophthalmic pathology with single variables previously measured in the Framingham study. *Am J Epidemiol* 1977; 106:33–41.

92. Barnett PA, Gonzalez RG, Chylack LT, et al: The effect of oxidation on sorbitol pathway kinetics. *Diabetes* 1986; 35:426-432.

93. Kuwabara T, Cogan DG: Retinal vascular patterns. VI. Mural cells of the retinal capillaries. *Arch Ophthalmol* 1963; 69:492-502.

94. Hogan MJ, Feeney L: The ultrastructure of the retinal vessels. II. The small vessels. *J Ultrastruct Res* 1963; 9:29-46.

95. Ishikawa T: Fine structure of retinal vessels in man and the macaque monkey. *Invest Ophthalmol* 1963; 2:1-15.

96. Shakib M, Cunha-Vaz JG: Studies on the permeability of the blood retinal barrier. IV. Junctional complexes of the retinal vessels and their role in the permeability of the blood retinal barrier. *Exp Eye Res* 1966; 5:229-234.

97. Speiser P, Gittelsohn AM, Patz A: Studies on diabetic retinopathy. III. Influence of diabetes on intramural pericytes. *Arch Ophthalmol* 1968; 80:332-337.

98. Yanoff M: Diabetic retinopathy. *N Engl J Med* 1966; 274:1344-1349.

99. Addison DJ, Garner A, Ashton N: Degeneration of intramural pericytes in diabetic retinopathy. *Br Med J* 1970; 1:264-266.

100. Cogan DG, Toussaint D, Kuwabara T: Retinal vascular patterns. IV. Diabetic retinopathy. *Arch Ophthalmol* 1976; 66:366-372.

101. Ashton N: The blood-retinal barrier and vaso-glial relationships in retinal disease. *Trans Ophthalmol Soc UK* 1965; 85:199-230.

102. Cunha-Vaz J, DeAbreau JRF, Campos AJ, et al: Early breakdown of the blood-retinal barrier in diabetes. *Br J Ophthalmol* 1975; 59:649-656.

103. Waltman S, Krupin T, Hanish S, et al: Alteration of the blood retinal barrier in experimental diabetes mellitus. *Arch Ophthalmol* 1978; 96:878-879.

104. Wise GN, Dollery GT, Henkind P: *The Retinal Circulation*, pp 290-324, 350-420. New York, Harper & Row, 1971.

105. Gabbay KH: Purification and immunological identification of bovine retinal aldose reductase. *Isr J Med Sci* 1972; 8:1626-1628.

106. Akagi Y, Kador PF, Kuwabara T, et al: Aldose reductase localization in human retinal mural cells. *Invest Ophthalmol Vis Sci* 1983; 24:1516-1519.

107. Ludvigson MA, Sorenson RL: Immunohistochemical localization of aldose reductase. II. Rat eye and kidney. *Diabetes* 1980; 29:450-459.

108. Kern TS, Engerman RL: Distribution of aldose reductase in ocular tissues. *Exp Eye Res* 1981; 33:175-182.

109. Akagi Y, Yajima Y, Kador PF, et al: Localization of aldose reductase in the human eye. *Diabetes* 1984; 33:562-566.

110. Buzney SM, Frank RN, Varma SD, et al: Aldose reductase in retinal mural cells. *Invest Ophthalmol Vis Sci* 1977; 16:392-396.

111. Kern TS, Engerman RL: Hexitol production by canine retinal microvessels. *Invest Ophthalmol Vis Sci* 1985; 26:382-384.

112. Russell P, Merola LO, Yajima Y, et al: Aldose reductase activity in a cultured human retinal cell line. *Exp Eye Res* 1982; 35:331-336.

113. Li W, Chan LS, Khatami M, et al: Characterization of glucose transport by bovine retinal capillary pericytes in culture. *Exp Eye Res* 1985; 41:191–199.

114. Hutton JC, Schofield PJ, Williams JF, et al: Sorbitol metabolism in the retina: Accumulation of pathway intermediates in streptozotocin induced diabetes in the rat. *Aust J Exp Biol Med Sci* 1974; 52:361–373.

115. MacGregor LC, Rosecan LR, Laties AM, et al: Microanalysis of total lipid, glucose, sorbitol, and *myo*-inositol in individual retinal layers of normal and alloxan diabetic rabbits. *Diabetes* 1984; 33:89A.

116. Li W, Shen S, Khatami M, et al: Stimulation of retinal capillary pericyte protein and collagen synthesis in culture by high-glucose concentration. *Diabetes* 1984; 33:785–789.

117. Engerman RL, Kern TS: Experimental galactosemia produces diabetic-like retinopathy. *Diabetes* 1984; 33:97–100.

118. Robison WG, Kador PF, Kinoshita JH: Retinal capillaries: Basement membrane thickening by galactosemia prevented with aldose reductase inhibitor. *Science* 1983; 221:1177–1179.

119. Li W, Khatami M, Rockey JH: The effects of glucose and an aldose reductase inhibitor on the sorbitol content and collagen synthesis of bovine retinal capillary pericytes in culture. *Exp Eye Res* 1985; 40:439–444.

120. Hotta N, Kakuta H, Fukasawa H, et al: Aldose reductase inhibitor and fructose-rich diet: Its effect on the development of diabetic retinopathy. *Diabetes* 1984; 33(suppl 1):199A.

121. Heath H, Hamlett YC: The sorbitol pathway: Effect of streptozotocin induced diabetes and the feeding of a sucrose-rich diet on glucose, sorbitol and fructose in the retina, blood and liver of rats. *Diabetologia* 1976; 12:43–46.

122. Kennedy A, Frank RN, Varma SD: Aldose reductase activity in retinal and cerebral microvessels and cultured vascular cells. *Invest Ophthalmol Vis Sci* 1983; 24:1250–1258.

123. MacGregor LC, Matschinsky FM: Treatment with aldose reductase inhibitor or with *myo*-inositol arrests deterioration of the electroretinogram of diabetic rats. *J Clin Invest* 1985; 76:887–889.

124. MacGregor LC, Matschinsky FM: Correlation of biochemical and electrophysiological abnormalities in retinas of experimentally diabetic animals. *Diabetes* 1985; 34(suppl 1):13A.

125. Yamani T: Effect of aldose reductase inhibitor on the oscillatory potential in ERG of streptozotocin diabetic rats. *Folia Ophthalmol Jpn* 1983; 34:2237–2244.

126. Li W, Chan S, Khatami M, et al: Inhibition of *myo*-inositol uptake in cultured bovine retinal capillary pericytes by D-glucose: Reversal by Sorbinil. *Invest Ophthalmol Vis Sci* 1985; 26(Suppl):335.

127. Krupin T, Waltman SR, Oestrich C, et al: Vitreous fluorophotometry in juvenile-onset diabetes mellitus. *Arch Ophthalmol* 1978; 96:812–814.

128. Krupin T, Waltman SR, Szewczyk P, et al: Fluorometric studies on the blood retinal barrier in experimental animals. *Arch Ophthalmol* 1982; 100:631–634.

129. Cunha-Vaz J, Mota C, Leite E, et al: Effect of aldose reductase inhibitors on the blood-retinal barrier in early diabetic retinopathy. *Diabetes* 1985; 34(suppl 1):109A.

130. Cunha-Vaz JG, Mota CC, Leite EC, et al: Effect of sulindac on the permeability of the blood retinal barrier in early diabetic retinopathy. *Arch Ophthalmol* 1985; 103:1307–1311.

131. Cunha-Vaz JG, Mota CC, Leite EC, et al: Effect of Sorbinil on blood-retinal barrier in early diabetic retinopathy. *Diabetes* 1986; 35:574–578.

132. Heath H, Kang SS, Philippou D: Glucose, glucose-6-phosphate, lactate and pyruvate content of the retina, blood and liver of streptozotocin-diabetic rats fed sucrose- or starch-rich diets. *Diabetologia* 1975; 11:57–62.

133. Hamlett YC, Heath H: The accumulation of fructose-1-phosphate in the diabetic rat retina. *IRCS Med Sci* 1977; 5:510.

134. Boot-Hanford R, Heath H: The effects of aldose reductase inhibitors on the metabolism of cultured monkey kidney epithelial cells. *Biochem Pharmacol* 1981; 30:3065–3069.

135. Poulsom R, Mirrlees DJ, Earl DCN, et al: The effects of an aldose reductase inhibitor upon the sorbitol pathway, fructose-1-phosphate and lactate in the retina and nerve of streptozotocin-diabetic rats. *Exp Eye Res* 1983; 36:751–760.

136. Schultz RO, VanHorn DL, Peters MA, et al: Diabetic keratopathy. *Trans Am Ophthalmol Soc* 1981; 79:180–199.

137. Hyndiuk RA, Kazarian EL, Schultz RO, et al: Neurotrophic corneal ulcers in diabetes mellitus. *Arch Ophthalmol* 1977; 95:2193–2196.

138. Pfister RR, Schepens CL, Lemp MA, et al: Photocoagulation keratopathy: Report of a case. *Arch Ophthalmol* 1971; 86:94–96.

139. Kanski JJ: Anterior segment complications of retinal photocoagulation. *Am J Ophthalmol* 1975; 79:424–427.

140. Perry HD, Foulks GN, Thoft RA, et al: Corneal complications after closed vitrectomy through the pars plana. *Arch Ophthalmol* 1978; 96:1401–1403.

141. Brightbill FS, Myers FL, Bresnick GN: Postvitrectomy keratopathy. *Am J Ophthalmol* 1978; 85:651–655.

142. Foulks GN, Thoft RA, Perry HD, et al: Factors related to corneal epithelial complications after closed vitrectomy in diabetes. *Arch Ophthalmol* 1979; 97:1076–1078.

143. Blankenship GW, Machemer R: Pars plana vitrectomy for the management of severe diabetic retinopathy: An analysis of results five years following surgery. *Ophthalmology (Rochester)* 1978; 85:553–559.

144. Faulborn J, Conway BP, Machemer R: Surgical complications of pars plana vitreous surgery. *Ophthalmology (Rochester)* 1978; 85:116–125.

145. Cogan DG, Kinoshita JH, Kador PF, et al: Aldose reductase and complications of diabetes. *Ann Intern Med* 1984; 101:82–91.

146. Akagi Y, Yajima Y, Kador PF, et al: Localization of aldose reductase in the human eye. *Diabetes* 1984; 33:562–566.

147. Friend J, Snip RC, Kiorpes TC, et al: Insulin insensitivity and sorbitol production of the normal rabbit corneal epithelium in vitro. *Invest Ophthalmol Vis Sci* 1980; 19:913–919.

148. Fukushi A, Merola LO, Tanaka M, et al: Re-epithelialization of denuded corneas in diabetic rats. *Exp Eye Res* 1980; 31:611–621.

149. Datiles MD, Kador PF, Fukui HN, et al: Corneal re-epithelialization in galactosemic rats. *Invest Ophthalmol Vis Sci* 1983; 24:563–569.

CHAPTER 4

Diabetic Neuropathy

Among the several mechanisms that have been proposed to explain the pathogenesis of diabetic neuropathy, the scheme invoking the polyol pathway as contributory to the nerve dysfunction in diabetes is one that has perhaps gained the most credence in recent years. In part this relates to a better understanding of the morphologic, functional, and metabolic abnormalities that underlie the neuropathic syndromes, and in part to the impetus provided by the pharmaceutical industry with premarketing publicity for the clinical use of aldose reductase inhibitors in the treatment of diabetic neuropathy.

Although the clinical manifestations of diabetic neuropathy may vary, the electrophysiologic and morphologic features reflecting neural involvement are fairly constant. Slowing of the motor nerve conduction velocity is the most traditionally recognized parameter of peripheral neuropathy, but sensory disturbances and autonomic nerve dysfunction can also be quantitated by techniques such as measurement of the sensory threshold, sensory nerve conduction velocities, and microneurography.[1-5] Histologically, there is segmental demyelination, loss of Schwann cells, and axonal shrinkage and degeneration.[6-9] Similar electrophysiologic and morphologic changes have been repeatedly observed in peripheral nerves of animals with experimental diabetes.[10-18]

The capillaries of the endoneurium of peripheral nerves in diabetic subjects display abnormalities reminiscent of those characteristically found in other tissues subject to diabetic complications, notably the retinal and glomerular microvasculature. These include thickening and reduplication of the capillary basement membrane, endothelial hyperplasia, and fibrin or platelet plugging.[19-22] Such abnormalities are believed to be associated with distal polyneuropathic syndromes, and should not be construed as tantamount to occlusive atherosclerotic lesions, nor confused with the ischemic vascular insults now recognized to be causally associated with mononeuropathic disease[23-25] (Table 4-1). However, the questions of how or if they contribute to, or if they result from, the primary pathogenetic process responsible for diabetic polyneuropathy remain enigmatic. Thickening of the perineurial cell basement membrane of the sural nerve and the dorsal root ganglion is also found in human diabetes.[26,27] These changes may arise from increased vascular permeability, perhaps related to the foregoing capillary changes, although some doubt has been cast on this hypothesis.[27,28]

The notion that increased activity of the polyol pathway might be involved in the pathogenesis of diabetic neuropathy was bolstered by the finding that sorbitol and fructose are present in high concentration in normal peripheral nerve,[29-32] and the observation that these substances further accumulate in the nerves of animals with experimental diabetes.[33-37] Sorbitol and fructose concentrations are also increased in peripheral nerves obtained from patients with diabetes,[8,38] and the level of sorbitol in cerebrospinal fluid of diabetic

TABLE 4-1   Etiologic Classification of Diabetic Neuropathies

Etiology	Syndrome
Ischemia	Mononeuropathy (e.g., cranial, radiculopathy)
Metabolic abnormalities	Distal symmetrical polyneuropathy
Ischemia plus metabolic abnormalities	Diabetic amyotrophy Diabetic neuropathic cachexia

subjects is higher than that in nondiabetic individuals.[39,40] Nerve is one of the tissues that do not require insulin for glucose transport,[41,42] and hence the glucose concentration in Schwann cells and within axons mirrors that of the extracellular compartment. The rise in intracellular glucose accompanying hyperglycemia in the diabetic state would be expected to activate the polyol pathway, increasing the formation of sorbitol and fructose in the nerve. When hyperglycemia is controlled with insulin therapy, the elevated levels of sorbitol and fructose in the nerve fall significantly, confirming that hyperglycemia resulting from insulin deficiency is primary to their increase.[33,34,36]

Biochemical and immunohistochemical studies have consistently demonstrated the presence of aldose reductase in the Schwann cell.[29,43-47] The enzyme is localized to the cytoplasm of the cell body and is not present in the axon, the perineurial structures, or the endoneurial space. This is compatible with earlier studies that had suggested that aldose reductase was confined to the cell cytoplasm, since activity persisted in Wallerian degenerated nerves, a model in which axons are resorbed but Schwann cells persist and proliferate.[43,44]

The traditional explanation implicating the polyol pathway in the pathogenesis of diabetic neuropathy has invoked damage resulting from the osmotic effects of accumulated sorbitol. In this construct, the sorbitol-induced edema of the Schwann cell would cause cell lysis and, eventually, demyelination due to loss of the myelin-sustaining role of these cells. However, electron microscopic studies have not shown that Schwann cell edema exists in animals with experimental diabetes[12,18,48-50]; in fact, one often-cited study found that the cytoplasmic volume of the Schwann cell was decreased, not increased, in diabetic rats.[14] Although cellular edema in peripheral nerve biopsies from patients with recent-onset diabetes has been described,[19] the inability to consistently document Schwann cell edema in experimental diabetes has dislodged this hypothesis from its position of primacy. Nevertheless, the concept that osmotic influences resulting from sorbitol accumulation compromise nerve function remains viable from another perspective. This relates to changes observed in the endoneurial compartment of peripheral nerves obtained from diabetic and galactose-fed animals. In the latter model, edema in the endoneurial interstitium and increased endoneurial pressure have been noted.[51-54] Decreased nerve conduction velocity, along with galactitol accumulation, accompanies these changes.[55,56] Endoneurial swelling has also been noted in nerves from hyperglycemic animals, although not invariably.[14,57,58]

Although it has been suggested that dehydration due to severe hyperglycemia might obscure tissue edema,[59] such dehydration consequent to a hyperosmolar stimulus might itself be injurious. One pathogenetic scheme that has been advanced invokes dehydration of the axon, resulting from hyperosmolarity of the endoneurial fluid, to explain the axonal shrinkage that is a morphologic hallmark of diabetic neuropathy.[9,50,58] In contrast, another hypothesis implicates endoneurial swelling, with dilution of the normal hypertonicity of the endoneurial fluid, in the compromise of axonal function.[14,59] Although seemingly dichotomous, both postulates subscribe to the view that axonal damage rather than demyelination is the primary nerve fiber lesion in diabetic neuropathy. If this is so, changes in the Schwann cell, albeit the neural reservoir of aldose reductase activity, would ensue secondary to axonal damage. Since inhibition of aldose reductase decreases swelling of the sciatic nerve in galactose-fed rats[60] and reduces the sorbitol concentration and improves motor nerve conduction velocity in the sciatic nerve of streptozotocin-diabetic rats,[61] it is believed that there is a causal link between enhanced polyol pathway activity, depressed nerve function, and neural edema, even if edema does not occur in the Schwann cell itself.

Another manner in which enhanced polyol pathway activity may alter axonal metabolism pertains to its influences on nerve *myo*-inositol. This cyclic hexitol is a ubiquitous constituent of cells and a component of tissue phosphoinositides, which are believed to have an important role in neural function. It has been known for a number of years that diabetic patients have greatly exaggerated excretion of *myo*-inositol.[39,62-66] Although the kidney appears to play a major role in the regulation of plasma *myo*-inositol levels in normal subjects,[67] *myo*-inositol metabolism may be altered in diabetic individuals who do not have a significant compromise in renal function. Diabetic patients also exhibit intolerance to oral inositol, due perhaps in part to impaired utilization.[62,68] One site where uptake is diminished is the peripheral nerve, which can normally maintain its free *myo*-inositol content against a large concentration gradient.

TABLE 4-2 Restoration of Sciatic Nerve Conduction Velocity and *myo*-Inositol Content in Streptozotocin-Diabetic Rats with *myo*-Inositol Feeding

Experimental Group	Plasma Glucose (mg/dL)	Plasma *myo*-Inositol (μM)	Sciatic *myo*-Inositol (mmol/kg)	Conduction Velocity (m/s)
Control	< 200	26–30	2.76 ± 0.06	61.9 ± 0.6
Diabetic	600	24–38	2.27 ± 0.17*	51.0 ± 1.0*
Diabetic, 1% MI in diet	600	207–285	4.63 ± 0.14**	60.5 ± 0.6**

*Significantly different from control.
**Significantly different from untreated diabetic.
MI, *myo*-inositol.
Data from Greene et al.[74]

Myo-inositol content is significantly decreased in the peripheral nerves of rats with experimental diabetes[33,69,70]; decreased *myo*-inositol in peripheral nerves obtained from human patients with diabetes has been found by some[37] but not all[8,35] investigators, and the *myo*-inositol content of cerebrospinal fluid in patients with neuropathy is less than that in patients without neuropathy.[71,72] The reduced peripheral nerve *myo*-inositol in experimental diabetes is associated with slowing of the conduction velocity.[25,63] Treatment with *myo*-inositol prevents both nerve *myo*-inositol depletion and the reduction in nerve conduction velocity[33,73-75] (Table 4-2). The corrective effect of oral *myo*-inositol supplementation occurs despite persistent severe hyperglycemia, whereas insulin therapy that does not adequately control hyperglycemia fails to restore either nerve *myo*-inositol levels or the depressed conduction velocity (Table 4-3). These findings firmly implicate hyperglycemia and/or insulin deficiency in the genesis of nerve *myo*-inositol depletion, and the reduction in nerve *myo*-inositol content in the genesis of the slowed nerve conduction velocity in diabetes.

A link between these findings and the polyol pathway was provided by the demonstration that inhibition of aldose reductase prevents not only the increase in nerve sorbitol but also the fall in nerve *myo*-inositol and the reduction in nerve conduction velocity.[75-78] Although the mechanism by which inhibition of one of the enzymes of the polyol pathway mediates this effect is not clear, the fact that an aldose reductase inhibitor can do so opens new windows

TABLE 4-3   Effect of Insulin Treatment on Sciatic Nerve Conduction Velocity and
myo-Inositol Content

Experimental Group	Conduction Velocity (m/s)	*myo*-Inositol Content (mmol/kg)
Control	64.6 ± 0.9	3.17 ± 0.25
Diabetic, untreated	50.1 ± 0.9*	2.19 ± 0.11*
Diabetic, insulin treated, inadequate control	51.4 ± 1.1	2.52 ± 0.13
Diabetic, insulin treated, meticulous control	62.9 ± 0.9**	2.98 ± 0.22**

*Significantly different from control.
**Significantly different from untreated diabetic.
Data from Greene et al.[33]

concerning the undoubtedly complex relationship among activation of the polyol pathway, altered *myo*-inositol metabolism, and the pathogenesis of diabetic neuropathy. It also renders imperfect, if not obsolete, an earlier explanation that had been put forward to explain the association of hyperglycemia with nerve *myo*-inositol depletion. This postulate developed from the observation that glucose competitively inhibits in vitro *myo*-inositol uptake in endoneurial preparations derived from rabbit sciatic nerve[79] (Table 4-4). In this preparation, more than 90% of the *myo*-inositol uptake occurs by a transport system that is sodium dependent and saturable, and the proposed hypothesis suggested that the hyperglycemia of diabetes inhibits the sodium-dependent *myo*-inositol uptake and is thereby responsible for the reduction in nerve *myo*-inositol content. Since inhibition of aldose reductase corrects *myo*-inositol depletion without altering the glucose concentration, it is clear that this construct is incomplete. One possibility is that glucose-induced polyol accumulation increases *myo*-inositol efflux from the cell. Support for this hypothesis derives from studies with lenses, in which increased *myo*-inositol efflux has been demonstrated when incubations are performed in media containing high glucose concentrations.[80-82]

The next link in the scheme relating altered polyol and *myo*-inositol metabolism to compromised nerve function in diabetes pertains to the recently described interrelationships of these substances

TABLE 4-4   Glucose Inhibition of Sodium-Dependent Uptake of [³H]-Labeled *myo*-Inositol by Rabbit Sciatic Nerve Endoneurial Preparations

Incubation Conditions	Uptake (nmol/kg/min)
5 µM *myo*-inositol	
5 mM glucose	78 ± 4
20 mM glucose	57 ± 3*
50 µM *myo*-inositol	
5 mM glucose	415 ± 43
20 mM glucose	310 ± 15*

*Significantly different from 5 mM glucose.
Data from Greene and Lattimer.[79]

with the activity of sodium/potassium adenosine triphosphatase (Na/K-ATPase), an enzyme critically involved with maintenance of sodium concentrations in the axon. Na/K-ATPase activity is reduced in the sciatic nerve of rats with experimental diabetes, and this abnormality is believed to cause an increase in the intra-axonal sodium concentration, resulting in impaired sodium influx during depolarization and a consequent compromise in the propagation of nerve impulses.[83,84] Decreased neural Na/K-ATPase activity is corrected by either dietary *myo*-inositol supplementation or treatment with the aldose reductase inhibitor sorbinil[85,86] (Table 4-5). These observations not only implicate *myo*-inositol depletion in the genesis of impaired Na/K-ATPase activity, but also indicate that enhanced polyol pathway activity, perhaps via reduction in cell *myo*-inositol levels, contributes to the defect in neural Na/K-ATPase activity in diabetes. Nerve conduction velocity, *myo*-inositol content, and Na/K-ATPase activity also are decreased in Bio-Breeding diabetic (BB/D) rats. Vigorous treatment with insulin normalizes these parameters and also prevents the progression of associated structural changes.[87] One such structural change consists of a decrease in the number of intramembranous particles in replicas of freeze fracture surfaces of the internodal myelin membranes of large sciatic nerve fibers.[88] This change, which may represent the structural correlate of the reported decrease in electrical resistance of internodal myelin in diabetes, is prevented by treatment with insulin or by dietary *myo*-inositol supplementation.

TABLE 4-5   Effect of Diabetes and Sorbinil Treatment on
Rat Sciatic Nerve ATPase Activity

Experimental Group	Na/K-ATPase (μmol/g/hr)
Control	104.3 ± 4.4
Untreated diabetic	56.8 ± 2.9*
Sorbinil-treated diabetic	95.7 ± 5.0**

*Significantly different from control.
**Significantly different from untreated diabetic.
Data from Greene and Lattimer.[86]

Na/K-ATPase is a membrane-bound enzyme complex intimately
associated with phospholipid-related cell cycles and components,
particularly phosphatidylinositol. This inositol-containing phospholipid is both a precursor to synthesis and a breakdown product of
the polyphosphoinositides, which are present in high concentration
in the peripheral nerve, where they are associated with myelin.
Phosphoinositide turnover is markedly increased during neural
stimulation, impulse generation, and synaptic transmission, and
may be involved with the generation of ionic fluxes that accompany
these processes.[89-94] Thus, diminished *myo*-inositol content may
compromise phosphoinositide synthesis or metabolism, in turn
impairing neuronal Na/K-ATPase activity and contributing to faulty
neuronal transmission. This could create a self-perpetuating cycle of
negative influences on nerve function, as Green[83,86,95] has proposed
(Figure 4-1).

One of the problems with this construct is that corresponding
quantitative changes in the phosphatidylinositol or polyphosphoinositide content of peripheral nerve in diabetes have not been
definitively established.[70,73,96,97] However, several studies have
shown that the activities of enzymes concerned with phosphatidylinositol and polyphosphoinositide synthesis are decreased in preparations of nerves obtained from diabetic animals.[69,97-99] Additionally, the in vitro incorporation of [32P]orthophosphate into sciatic
nerve polyphosphoinositides, particularly phosphatidylinositol-
4,5-bisphosphate (PIP$_2$), is increased in preparations from diabetic
rats, and these increases are prevented by treatment with insulin.[100,101] The finding that the glucose-induced inhibition of *myo*-

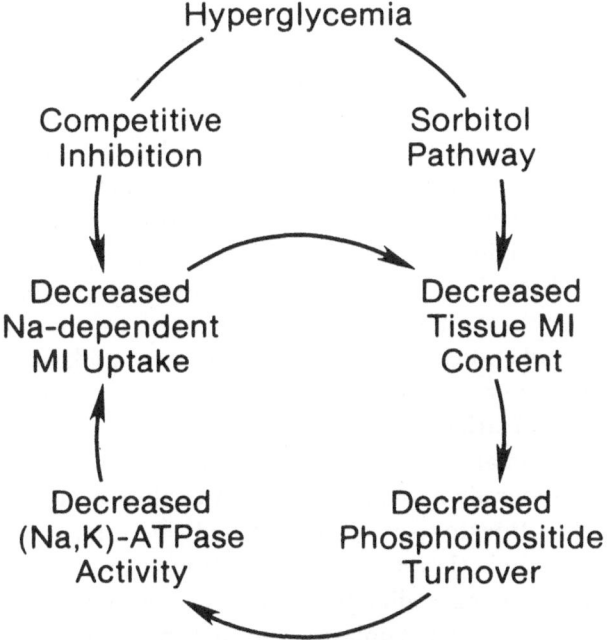

FIGURE 4-1 Proposed interaction of hyperglycemia, sodium-dependent *myo*-inositol (MI) uptake, inositol phospholipid metabolism, and sodium-potassium ATPase [(Na,K)-ATPase] activity in the pathogenesis of diabetic neuropathy. Reproduced with permission from the American Diabetes Association, Inc. From Greene DA, Lattimer SA: Protein kinase C agonists acutely normalize decreased ouabain-inhibitable respiration in diabetic rat nerve. *Diabetes* 1986; 35:242–245.

inositol uptake in adipocytes is associated with decreased phosphatidylinositol formation also offers support for this hypothesis.[102] Finally, Greene and Lattimer recently reported that agonists of protein kinase C, a putative messenger regulated by phosphoinositide turnover, normalize the reduced oxygen consumption that reflects diminished Na/K-ATPase activity in endoneurial preparations from alloxan-diabetic rabbits.[95] This suggests that nerve *myo*-inositol depletion impairs Na/K-ATPase activity by a mechanism involving protein kinase C, and that altered phosphoinositide metabolism, presumably related to *myo*-inositol deficiency, underlies the defect in protein kinase C. Other evidence linking phosphoinositide turnover, Na/K-ATPase activity, and *myo*-inositol metabolism has been found in aortic tissue (see Chapter 6), and an analogous sequence of metabolic abnormalities has been identified in peripheral nerve. In rabbit tibial nerve endoneurial preparations, deprivation of *myo*-inositol results in associated decreases in resting energy utilization, Na/K-ATPase activity, and phosphatidylinositol metabolism in a compartment of rapid basal turnover.[103,104]

Recently, defects in axonal transport, known for several years to occur in peripheral nerve of animals with experimental diabetes, have been linked to the accumulation of polyol pathway products.[77,105-113] Axonal transport is assessed by measuring the accumulation of endogenous components of the axoplasm after constriction of the nerve trunk; accumulation in nerve segments proximal to the constriction reflects orthograde transport, whereas accumulation distal to the ligature is an index of retrograde transport. The orthograde axonal transport of choline acetyltransferase and acetylcholinesterase is diminished in sciatic nerves of streptozotocin-diabetic rats. The implication of this abnormality is that there is restricted availability of neuronal proteins and nutrients to distal portions of the axoplasm, and that this restriction contributes to the axonopathy of diabetic neuropathy. Decreased conduction velocity accompanies impaired axonal transport, and insulin treatment normalizes both parameters. More importantly, from the perspective of the interrelationships between changes in this axonal function and alterations in polyol and/or *myo*-inositol metabolism, treatment with the aldose reductase inhibitor Sorbinil or with dietary *myo*-inositol supplementation restores axonal transport in streptozotocin-diabetic rats. Since either of these treatments also restores nerve *myo*-inositol levels, *myo*-inositol depletion is believed to underlie this defect in axonal transport. However, nerve *myo*-inositol levels were not decreased in diabetic BB/D rats that were treated with insulin but remained hyperglycemic, and choline acetyltransferase did not accumulate after constriction of the sciatic nerve in these animals. Although differences in the nature of the diabetes (experimental versus spontaneous) in the animals used for these two studies may account for the seemingly dichotomous results, the findings in the latter study are compatible with the interpretation that *myo*-inositol depletion is prerequisite to this impairment in axonal transport. On the other hand, neither aldose reductase inhibition nor *myo*-inositol supplementation prevents the deficit in the slow anterograde transport of labeled protein in motoneurons of rats with diabetes of 4 weeks' duration.[114,115]

The aldose reductase inhibitor Statil, developed by Imperial Chemical Industries (ICI 128436), prevented sorbitol accumulation, *myo*-inositol depletion, and the defective accumulation of choline acetyltransferase activity occurring after ligature in the vagus nerve of streptozotocin-diabetic rats.[116] Choline acetyltransferase activity in cholinergically innervated organs such as the terminal ileum and iris is reduced in streptozotocin diabetes, and treatment with Statil prevents these reductions.[117] *Myo*-inositol levels and Na/K-ATPase activity are reduced in the superior cervical ganglia of diabetic rats, and these changes are prevented by treatment with Sorbinil.[118] The implication of these findings is that the mechanisms underlying autonomic neuropathy are similar to those responsible for peripheral neuropathy, at least in the experimentally diabetic rat.

At least three interrelated and fundamental concerns have limited the translation of the intriguing results from these studies with tissue from diabetic animals to the problem of diabetic neuropathy in humans and its possible treatment with aldose reductase inhibitors. First, there is no assurance that the biochemical abnormalities associated with defects in nerve function in animal models of diabetes are the same as those responsible for the development of diabetic neuropathy in people. Second, many of the studies have employed measurement of motor nerve conduction velocity to detect the appearance of functional deficits or restoration of functional integrity by various manipulations, yet nerve conduction reflects function only in that small fraction of total nerve fibers that are the largest and most rapidly conducting. Further, decreased nerve conduction velocity can occur in the absence of structural changes such as segmental demyelination and axonal loss. Third, in view of the fact that functional defects in diabetic neuropathy are associated with structural as well as metabolic abnormalities, there is no guarantee that agents having a favorable impact on dysfunction related to metabolic disturbances will also alleviate abnormalities arising from structural changes. Indeed, available evidence suggests dissociation of the underlying structural and metabolic factors in chronic diabetes. After 12 weeks of untreated diabetes followed by 6 weeks of vigorous treatment with insulin, sural nerve conduction velocity in BB rats was only partially restored despite normalization of nerve *myo*-inositol and Na/K-ATPase activity.[119] The residual defect in nerve conduction was associated with a persistent morphologic change consisting of a disappearance of paranodal axoglial junctional complexes. This abnormality has also been identified in the sural nerve of human patients with distal symmetrical diabetic polyneuropathy.[120] It is not yet known whether some of the structural changes found in experimental diabetes of lesser duration are reversible. Such changes would include the expanded size and number of filaments in the proximal axons and the shrinkage in cross-sectional area of distal axons of the sciatic nerve system in rats with streptozotocin diabetes of 4 to 6 weeks' duration.[121] Paranodal nerve fiber swelling, believed to be an ultrastructural precursor of axoglial dysjunction, is prevented by treatment with insulin or oral *myo*-inositol.[122]

References

1. Chochinov RH, Ullyot LE, Moorhouse JA: Sensory perception thresholds in patients with juvenile diabetes and their close relatives. *N Engl J Med* 1972; 286:1233–1237.

2. Graf RJ, Halter JB, Halar E, et al: Nerve conduction abnormalities in untreated maturity-onset diabetes: Relation to levels of fasting plasma glucose and glycosylated hemoglobin. *Ann Intern Med* 1979; 90:298–303.

3. Pfeifer MA, Weinberg CR, Cook DL, et al: Correlations among autonomic, sensory and motor neural function tests in untreated non-insulin-dependent diabetic individuals. *Diabetes Care* 1985; 8:576–584.

4. Valbo AB, Hagbarth K-E, Torebjork HE, et al: Somatosensory, proprioceptive and sympathetic activity in human peripheral nerves. *Physiol Rev* 1979; 59:919–957.

5. Fagius J: Microneurographic findings in diabetic polyneuropathy with special reference to sympathetic nerve activity. *Diabetologia* 1982; 23:415–420.

6. Thomas PK, Lascelles RG: The pathology of diabetic neuropathy. *Q J Med* 1966; 24:489–509.

7. Behse F, Buchthal F, Carlson FI: Nerve biopsy and conduction studies in diabetic neuropathy. *J Neurol Neurosurg Psychiatry* 1977; 40:1072–1082.

8. Dyck PJ, Sherman WR, Hallcher LM, et al: Human diabetic endoneurial sorbitol, fructose and *myo*-inositol related to sural nerve morphometry. *Ann Neurol* 1980; 8:590–596.

9. Clements RS, Bell DHS: Diagnostic, pathogenetic and therapeutic aspects of diabetic neuropathy. *Spec Top Endocrinol Metab* 1982; 3:1–43.

10. Powell H, Know D, Lee S, et al: Alloxan diabetic neuropathy: Electron microscopic studies. *Neurology* 1977; 27:60–66.

11. Schlaepfer WW, Gerritsen GC, Dulin WE: Segmental demyelination in the distal peripheral nerves of chronically diabetic chinese hamsters. *Diabetologia* 1974; 10:541–548.

12. Yagihashi S, Kudo K, Nishihira M: Peripheral nerve structures of experimental diabetes in rats and the effect of insulin treatment. *Tohoku J Exp Med* 1979; 127:35–44.

13. Jakobsen J: Axon dwindling in early experimental diabetes. I. A study of cross-sectioned nerves. *Diabetologia* 1976; 12:539–546.

14. Jakobsen J: Peripheral nerves in early experimental diabetes. Expansion of the endoneurial space as a cause of increased water content. *Diabetologia* 1978; 14:113–119.

15. Jakobsen J: Early and preventable changes of peripheral nerve structure and function in insulin-deficient diabetic rat. *J Neurol Neurosurg Psychiatry* 1979; 42:509–518.

16. Monckton G, Pehoevich E: Autonomic neuropathy in the streptozotocin diabetic rat. *Can J Neurol Sci* 1980; 7:135–142.

17. Moore SA, Peterson RG, Felten DL, et al: Reduced sensory and motor conduction velocity in 25-week-old diabetic [C57BL/Ks (db/db)] mice. *Exp Neurol* 1980; 70:548–555.

18. Sima AA: Peripheral neuropathy in the spontaneously diabetic BB-Wistar rat. *Acta Neuropathol* 1980; 51:223–227.

19. Bischoff A: Morphology of diabetic neuropathy. *Horm Metab Res* 1980; 9(suppl):18–28.

20. Timperley WR: Vascular and coagulation abnormalities in diabetic neuropathy and encephalopathy. *Horm Metab Res* 1980; 9(suppl):43–49.

21. Timperley WR, Ward JD, Preston FE, et al: Clinical and histological studies in diabetic neuropathy: A reassessment of vascular factors in relation to vascular coagulation. *Diabetologia* 1976; 12:237–243.

22. Williams E, Timperley WR, Ward JD, et al: Electron microscopic studies of vessels in diabetic peripheral neuropathy. *J Clin Pathol* 1980; 33:462–470.

23. Raff MC, Asbury AK: Ischemic mononeuropathy and mononeuropathy multiplex in diabetes mellitus. *N Engl J Med* 1968; 279:17–22.

24. Asbury AK, Aldridge H, Hershberg R, et al: Oculomotor palsy in diabetes mellitus: A clinico-pathologic study. *Brain* 1970; 93:555–566.

25. Dreyfus PM, Hakim S, Adams RD: Diabetic ophthalmoplegia: Report of a case with postmortem study and comments on vascular supply of human oculomotor nerve. *Arch Neurol Psychiatry* 1957; 77:337–349.

26. Johnson PC, Brendel K, Meezan E: Human diabetic perineurial cell basement membrane thickening. *Lab Invest* 1981; 44:165–170.

27. Johnson PC: Thickening of the human dorsal root ganglion perineurial cell basement membrane in diabetes mellitus. *Muscle Nerve* 1983; 6:561–565.

28. Seneviratne KN: Permeability of blood nerve barriers in the diabetic rat. *J Neurol Neurosurg Psychiatry* 1972; 35:156–162.

29. Gabbay KH, Merola LO, Field RA: Sorbitol pathway: Presence in nerve and cord with substrate accumulation in diabetes. *Science* 1966; 151:209–210.

30. Sherman WR, Stewart MA: Identification of sorbitol in mammalian nerve. *Biochem Biophys Res Commun* 1966; 22:492–497.

31. Stewart MA, Passoneau JV: Identification of fructose in mammalian nerve. *Biochem Biophys Res Commun* 1964; 17:536–541.

32. Stewart MA, Sherman WR, Anthony S: Free sugars in alloxan diabetic rat nerve. *Biochem Biophys Res Commun* 1966; 22:488–491.

33. Greene DA, DeJesus PV, Winegrad AI: Effects of insulin and dietary *myo*-inositol on impaired peripheral motor nerve conduction velocities in acute streptozotocin diabetes. *J Clin Invest* 1975; 55:1326–1336.

34. Stewart MA, Sherman WR, Kurien MM, et al: Polyol accumulation in nervous tissue of rats with experimental diabetes and galactosemia. *J Neurochem* 1977; 14:1057–1066.

35. Ward JD: The polyol pathway in the neuropathy of early diabetes, in Camerini-Davalos RA, Cole HS (eds): *Vascular and Neurologic Changes in Early Diabetes*, pp 525–532. New York, Academic Press, 1973.

36. Ward JD, Baker RWP, Davis BH: Effect of blood sugar control on the accumulation of sorbitol and fructose in nervous tissues. *Diabetes* 1972; 21:1173–1178.

37. Poulsom R, Heath H: Inhibition of aldose reductase in five tissues of the streptozotocin diabetic rat. *Biochem Pharmacol* 1983; 32:1495–1499.

38. Mayhew JA, Gillon KRW, Hawthorne JN: Free and lipid inositol and sugars in sciatic nerve obtained postmortem from diabetic patients and control subjects. *Diabetologia* 1983; 24:13–15.

39. Pitkänen E, Servo C: Cerebrospinal fluid polyols in patients with diabetes. *Clin Chim Acta* 1973; 44:437–442.

40. Servo C, Pitkänen E: Variation in polyol levels in cerebrospinal fluid and serum in diabetic patients. *Diabetologia* 1975; 11:575–580.

41. Greene DA, Winegrad AI: In vitro studies of the substrates for energy production and the effects of insulin on glucose utilization in the neural components of peripheral nerve. *Diabetes* 1979; 28:878–887.

42. Greene DA, Winegrad AI: Effects of acute experimental diabetes on composite energy metabolism in peripheral nerve axons and Schwann cells. *Diabetes* 1979; 30:967–974.

43. Gabbay KH, O'Sullivan JB: The sorbitol pathway in diabetes and galactosemia: Enzyme and substrate localization. *Diabetes* 1968; 17:239–243.

44. Gabbay KH: The sorbitol pathway and complications of diabetes. *N Engl J Med* 1973; 288:831–836.

45. Gabbay KH: Hyperglycemia, polyol metabolism and complications of diabetes. *Annu Rev Med* 1975; 26:521–536.

46. Ludvigson MA, Sorenson RL: Immunohistochemical localization of aldose reductase. I. Enzyme purification and antibody preparation-localization in peripheral nerve, artery, and testes. *Diabetes* 1980; 29:438–449.

47. Kern TS, Engerman RL: Immunohistochemical distribution of aldose reductase. *Histochem J* 1982; 14:507–515.

48. Brown MJ, Sumner AJ, Greene DA, et al: Distal neuropathy in experimental diabetes. *Ann Neurol* 1980; 8:168–178.

49. Hanker JS, Ambrose WW, Yates PE, et al: Peripheral neuropathy in mouse hereditary diabetes mellitus. I. Comparison of neurologic, histologic and morphometric parameters with dystonic mice. *Acta Neuropathol (Berl)* 1980; 51:145–153.

50. Sugimura K, Windebank AJ, Natarajan V, et al: Interstitial hyperosmolarity may cause axis cylinder shrinkage in streptozotocin diabetic nerve. *J Neuropathol Exp Neurol* 1980; 39:710–721.

51. Low PA, Dyck PJ, Schmelzer JD: Mammalian peripheral nerve sheath has unique responses to chronic elevations of endoneurial fluid pressure. *Exp Neurol* 1980; 70:300–306.

52. Low PA, Dyck PJ, Schmelzer JD: Chronic elevation of endoneurial fluid pressure is associated with low-grade fiber pathology. *Muscle Nerve* 1982; 5:162–165.

53. Myers RR, Costello ML, Powell HC: Increased endoneurial fluid pressure in galactose neuropathy. *Muscle Nerve* 1979; 2:299–303.

54. Powell HC, Myers RR: Schwann cell changes and demyelination in chronic galactose neuropathy. *Muscle Nerve* 1983; 6:218–227.

55. Sharma AK, Baker RWR, Thomas PK: Peripheral nerve abnormalities related to galactose administration in rats. *J Neurol Neuropathol Psychiatry* 1976; 39:794–802.

56. Gabbay KH, Snider JJ: Nerve conduction defect in galactose-fed rats. *Diabetes* 1972; 21:295–300.

57. Powell HC, Costello ML, Myers RR: Endoneurial fluid pressure in experimental models of diabetic neuropathy. *J Neuropathol Exp Neurol* 1981; 40:613–634.

58. Dyck PJ, Lamberg EH, Windebank AJ, et al: Acute hyperosmolar hyperglycemia causes axonal shrinkage and reduced nerve conduction velocity. *Exp Neurol* 1981; 71:507–514.

59. Powell HC: Pathology of diabetic neuropathy: New observations, new hypotheses. *Lab Invest* 1983; 49:515–518.

60. Cogan DG, Kinoshita JH, Kador PF, et al: Aldose reductase and complications of diabetes. *Ann Intern Med* 1984; 101:82–91.

61. Yue DK, Hanwell MA, Satchell PM, et al: The effect of aldose reductase inhibition on motor nerve conduction velocity in diabetic rats. *Diabetes* 1982; 31:789–794.

62. Clements RS, Reynertson RH, Starnes WS: *Myo*-inositol metabolism in diabetes mellitus. *Diabetes* 1974; 23:348A.

63. Daughaday WH, Larner J: The renal excretion of inositol in normal and diabetic human beings. *J Clin Invest* 1954; 33:326–332.

64. Clements RS: Diabetic neuropathy—Newer concepts in etiology. *Diabetes* 1979; 28:604–611.

65. Clements RS, Reynertson R: *Myo*-inositol metabolism in diabetes mellitus. Effect of insulin treatment. *Diabetes* 1977; 26:215–221.

66. Pitkanen E: The serum polyol pattern and the urinary polyol excretion in diabetic and uremic patients. *Clin Chim Acta* 1972; 38:221–230.

67. Clements RS, Diethelm AG: The metabolism of *myo*-inositol by the human kidney. *J Lab Clin Med* 1979; 93:210–219.

68. Barbosa J: Plasma *myo*-inositol in diabetics including patients with renal allografts. *Acta Diabetol Lat* 1978; 15:95–101.

69. Clements RS, Stockard CR: Abnormal sciatic nerve *myo*-inositol metabolism in the streptozotocin-diabetic rat. Effects of insulin treatment. *Diabetes* 1980; 29:227–235.

70. Palamano KP, Whiting PH, Hawthorne JH: Free and lipid *myo*-inositol in tissues from rats with acute and less severe streptozotocin-induced diabetes. *Biochem J* 1977; 167:229–235.

71. Servo C: Sorbitol and *myo*-inositol in the cerebrospinal fluid of diabetic patients. *Acta Endocrinol* 1980; 94(suppl 238):133–138.

72. Servo C, Bergstrom L, Fogelholm R: Cerebrospinal fluid sorbitol and *myo*-inositol in diabetic polyneuropathy. *Acta Med Scand* 1977; 202:301–304.

73. Winegrad AI, Greene DA: Diabetic polyneuropathy: The importance of insulin deficiency, hyperglycemia and alterations in *myo*-inositol metabolism in its pathogenesis. *N Engl J Med* 1976; 295:1416–1421.

74. Greene DA, Lewis RA, Lattimer SA, et al: Selective effects of *myo*-inositol administration on sciatic and tibial motor nerve conduction parameters in the streptozotocin-diabetic rat. *Diabetes* 1982; 31:573–578.

75. Gillon KRW, Hawthorne JN, Tomlinson DR: *Myo*-inositol and sorbitol metabolism in relation to peripheral nerve function in experimental diabetes in the rat: The effect of aldose reductase inhibition. *Diabetologia* 1983; 25:365–371.

76. Gillon KRW, Hawthorne JN: Sorbitol, inositol and nerve conduction in diabetes. *Life Sci* 1983; 32:1943–1947.

77. Mayer JH, Tomlinson DR: Prevention of defects of axonal transport and nerve conduction velocity by oral administration of *myo*-inositol or an aldose reductase inhibitor in streptozotocin-diabetic rats. *Diabetologia* 1983; 25:433–438.

78. Finegold D, Lattimer SA, Nolle S, et al: Polyol pathway activity and *myo*-inositol metabolism: A suggested relationship in the pathogenesis of diabetic neuropathy. *Diabetes* 1983; 32:988–992.

79. Greene D, Lattimer SA: Sodium and energy-dependent uptake of *myo*-inositol by rabbit peripheral nerve. *J Clin Invest* 1982; 70:1009–10018.

80. Cotlier E: *Myo*-inositol active transport by the crystalline lens. *Invest Ophthalmol* 1970; 9:681–691.

81. Chylack LT, Kinoshita JH: A biochemical evaluation of a cataract induced in a high-glucose medium. *Invest Ophthalmol* 1969; 8:401–412.

82. Broekhuyse RM: Changes in *myo*-inositol permeability in the lens due to cataractous conditions. *Biochim Biophys Acta* 1968; 163:269–272.

83. Greene DA, Lattimer SA: Impaired energy utilization and sodium-potassium ATPase in diabetic peripheral nerve. *Am J Physiol* 1984; 246:E311–E318.

84. Brismar T, Sima AAF: Changes in nodal function in nerve fibres of the spontaneously diabetic BB-Wistar rat: Potential clamp analysis. *Acta Physiol Scand* 1981; 113:499–506.

85. Greene DA, Lattimer SA: Impaired rat sciatic nerve sodium-potassium ATPase in acute streptozotocin diabetes and its correction by dietary *myo*-inositol supplementation. *J Clin Invest* 1983; 72:1058–1063.

86. Greene DA, Lattimer SA: Action of sorbinil in diabetic peripheral nerve. Relationships of polyol (sorbitol) pathway inhibition to a *myo*-inositol mediated defect in sodium-potassium ATPase activity. *Diabetes* 1984; 33:712–716.

87. Sima AAF, Lattimer SA, Yaghihashi S, et al: Biochemical, functional, and structural correction of diabetic neuropathy in the BB-rat after insulin treatment. *Fed Proc* 1984; 43:375A.

88. Fukuma M, Carpentier J-L, Orci L, et al: An alteration in internodal myelin membrane structure in large sciatic nerve fibres in rats with acute streptozotocin diabetes and impaired nerve conduction velocity. *Diabetologia* 1978; 15:65–72.

89. White GL, Larabee MG: Phosphoinositides and other phospholipids in sympathetic ganglia and nerve trunks of rats. *J Neurochem* 1973; 20:783–798.

90. White GL, Schellhase HU, Hawthorne JN: Phosphoinositide metabolism in rat superior cervical ganglion, vagus and phrenic nerve: Effects of electrical stimulation and various blocking agents. *J Neurochem* 1974; 28:149–158.

91. Hawthorne JN, Pickard MP, Griffin HD: Phosphatidylinositol, triphosphoinositide and synaptic transmission, in Wells WW, Eisenberg F (eds): *Cyclitols and Polyphosphoinositides*, pp 145–151. New York, Academic Press, 1978.

92. Henrickson HS, Reinertsen JL: Phosphoinositide interconversion: A model for control of Na^+ and K^+ permeability in the nerve axon. *Biochem Biophys Res Commun* 1971; 44:1258–1264.

93. Michell RH: Inositol phospholipids and cell surface receptor function. *Biochim Biophys Acta* 1980; 415:81–147.

94. Tetjak AG, Limarenko IM, Lossova GV, et al: Interrelation of phosphoinositide metabolism and ion transport in crab nerve fiber. *J Neurochem* 1977; 28:199–205.

95. Greene DA, Lattimer SA: Protein kinase C agonists acutely normalize decreased ouabain-inhibitable respiration in diabetic rat nerve. *Diabetes* 1986; 35:242–245.

96. Jeffreys JGR, Palmano KP, Sharma AK, et al: Influence of dietary *myo*-inositol on nerve conduction and inositol phospholipids in normal and diabetic rats. *J Neurol Neurosurg Psychiatry* 1978; 41:333–339.

97. Natarajan V, Dyck PJ, Schmid HO: Alterations in inositol lipid metabolism of rat sciatic nerve in streptozotocin-induced diabetes. *J Neurochem* 1981; 36:413–419.

98. Hothersall JS, McLean P: Effect of diabetes and insulin on phosphatidylinositol synthesis in rat sciatic nerve. *Biochem Biophys Res Commun* 1979; 88:477–484.

99. Whiting PH, Palmano KP, Hawthorne JN: Enzymes of *myo*-inositol and inositol lipid metabolism in rats with streptozotocin-induced diabetes. *Biochem J* 1979; 179:549–553.

100. Bell ME, Peterson RG, Eichberg J: Metabolism of phospholipids in peripheral nerve from rats with chronic streptozotocin-induced diabetes: Increased turnover of phosphatidyl-4,5-bisphosphate. *J Neurochem* 1982; 39:192–200.

101. Berti-Mattera L, Peterson R, Bell M, et al: Effect of hyperglycemia and its prevention by insulin treatment on the incorporation of [^{32}P] into polyphosphoinositides and other phospholipids in peripheral nerve of the streptozotocin diabetic rat. *J Neurochem* 1985; 45:1692–1698.

102. Kaplan SA, Lee W-NP, Scott ML: Glucose inhibits *myo*-inositol transport and phosphatidylinositol formation in adipocytes. *Diabetes* 1985; 34(suppl 1):183A.

103. Simmons DA, Winegrad AI, Martin DB: Significance of tissue *myo*-inositol concentrations in metabolic regulation in nerve. *Science* 1982; 217:848–851.

104. Winegrad AI, Simmons DA, Martin DB: Has one diabetic complication been explained? *N Engl J Med* 1983; 308:152–154.

105. Schmidt RE, Matschinsky FM, Godfrey DA, et al: Fast and slow axoplasmic flow in sciatic nerve of diabetic rats. *Diabetes* 1975; 24:1081–1085.

106. Giachetti A: Axoplasmol transport of noradenaline in the sciatic nerves of spontaneously diabetic mice. *Diabetologia* 1979; 16:191–199.

107. Jakobsen J, Sidenius P: Decreased axonal transport of structural proteins in streptozotocin-diabetic rats. *J Clin Invest* 1980; 66:292–297.

108. Tomlinson DR, Moriarty RJ, Mayer JH: Prevention and reversal of defective axonal transport and motor nerve conduction velocity in rats with experimental diabetes by treatment with the aldose reductase inhibitor Sorbinil. *Diabetes* 1984; 33:470–476.

109. Sidenius P: The axonapathy of diabetic neuropathy. *Diabetes* 1982; 31:356–363.

110. Tomlinson DR, Holmes PR, Mayer JH: Reversal, by treatment with an aldose reductase inhibitor, of impaired axonal transport and motor nerve conduction velocity in experimental diabetes. *Neurosci Lett* 1982; 31:189–193.

111. Mayer JH, Tomlinson DR: Axonal transport of cholinergic transmitter enzymes in vagus and sciatic nerves of rats with acute experimental diabetes mellitus; correlation with motor nerve conduction velocity and effects of insulin. *Neuroscience* 1983; 9:951–957.

112. Mayer JH, Herberg L, Tomlinson DR: Axonal transport and nerve conduction and their relation to nerve polyol and *myo*-inositol levels in spontaneously diabetic BB/D rats. *Neurochem Pathol* 1984; 2:285–293.

113. Tomlinson DR, Mayer H: Defects of axonal transport in diabetes mellitus—a possible contribution to the aetiology of diabetic neuropathy. *J Auton Pharmacol* 1984; 4:59–72.

114. Mayer JH, Tomlinson DR, McLean WG: Slow orthograde axonal transport of radiolabelled protein in sciatic motoneurones of rats with short-term experimental diabetes: Effects of treatment with an aldose reductase inhibitor of *myo*-inositol. *J Neurochem* 1984; 43:1265–1270.

115. Tomlinson DR, Sidenius P, Larsen JR: Slow component-a of axonal transport, nerve *myo*-inositol, and aldose reductase inhibition in diabetic rats. *Diabetes* 1986; 35:398–402.

116. Tomlinson DR, Townsend J, Fretten P: Prevention of defective axonal transport in streptozotocin-diabetic rats by treatment with "Statil" (ICI 128436), an aldose reductase inhibitor. *Diabetes* 1985; 34:970–972.

117. Tomlinson DR, Townsend S: Protection by the aldose reductase inhibitor "Statil" (ICI 128436) against loss of an axonally transported enzyme in nerve terminals of diabetic rats. *Diabetes* 1985; 34(suppl 1):202A.

118. Wells AM, Greene DA: A Sorbinil-responsive *myo*-inositol-related Na/K-ATPase defect in diabetic rat superior cervical ganglion. *Diabetes* 1985; 34(suppl 1):102A.

119. Sima AF, Lattimer SA, Yagihashi S, et al: Axo-glial dysjunction. A novel structural lesion that accounts for poorly reversible slowing of nerve conduction in the spontaneously diabetic Bio-Breeding rat. *J Clin Invest* 1986; 77:424–484.

120. Sima AAF, Bril V, Greene DA: A new characteristic ultrastructural abnormality, and morphologic evidence for pathogenetic heterogeneity in human diabetic neuropathy. *Clin Res* 1986; 34:688A.

121. Medori R, Autilio-Gambetti L, Monaco S, et al: Experimental diabetic neuropathy: Impairment of slow transport with changes in axon cross-sectional area. *Proc Natl Acad Sci USA* 1985; 82:7716–7720.

122. Greene DA, Lattimer SA, Sima AAF: Acute paranodal nerve fiber swelling and conduction slowing in the insulin-deficient BB rat reflects *myo*-inositol depletion and Na/K-ATPase deficiency rather than sorbitol accumulation. *Clin Res* 1986; 34:683A.

CHAPTER 5

Diabetic Nephropathy

For many years after identification of the polyol pathway and the implication that it participated in the pathogenesis of certain complications of diabetes, evidence that it might play a role in the development of diabetic nephropathy was largely inferential. Fundamental to the support of such a hypothesis would be the demonstration that aldose reductase activity and sorbitol content are increased in the characteristic tissue sites of the diabetic renal lesions. If there is a causal link between polyol accumulation and diabetic nephropathy, then aldose reductase activity should be demonstrable in glomerular tissue, and the glomerular polyol content should be elevated in diabetes.

Initial attempts to fulfill these criteria yielded negative experimental results. One early study found that neither of the sorbitol pathway intermediates, sorbitol and fructose, could be detected in glomeruli isolated from rats with streptozotocin diabetes left untreated for 14 days.[1] The same investigators reported that sorbitol and fructose did not accumulate in metabolically active glomeruli isolated from normal rats and incubated for 2 hours in media containing 590 mg% glucose.[1] Analysis by assay of enzyme activity in frozen dried tissue showed no change in the amount of sorbitol converted to glucose by glomeruli from rats with alloxan diabetes of 14 days' duration compared to control.[2] However, sorbitol dehydrogenase activity, also measured by enzymatic assay, was abundantly present in glomeruli and was increased in glomerular samples from alloxan-diabetic rats.[2] A subsequent study identified aldose reductase activity, measured as the conversion of sorbitol to glucose, in glomeruli from nondiabetic and diabetic human subjects[3]; aldose reductase activity was significantly elevated in diabetic samples, but sorbitol dehydrogenase activity was reduced in glomeruli from diabetic subjects (Table 5-1). Immunohistochemical study of frozen sections with an antibody raised against canine renal medullary aldose reductase demonstrated the presence of the enzyme in a wide variety of tissues in the dog, including several loci in the kidney, but not in renal glomeruli.[4] Several other studies identified aldose reductase in the renal medullae, papillae, and tubules[1,2,5] and in view of this it was proposed that sorbitol accumulation at such sites promoted osmotic swelling and thereby contributed to the tubular nephropathy that may be associated with uncontrolled diabetes.[6] Using specific antibodies to aldose reduc-

TABLE 5-1   Aldose Reductase and Sorbitol Dehydrogenase Activity in Human Glomeruli

Sample	Age (years)	Duration of Diabetes (years)	Activity (nmol/kg/hr)	
			Aldose Reductase	Sorbitol Dehydrogenase
Nondiabetic	19		22.8	18.6
	45		8.8	7.2
	56		9.1	8.7
	57		1.3	8.8
	65		12.0	11.4
	Mean ± SD		10.8±7.8	10.9±4.5
Diabetic	19	0.5	31.9	2.4
	37	5	76.8	5.4
	52	16	48.4	1.8
	68	13	11.6	3.3
	84	7	89.4	5.5
	Mean ± SD		51.6±31.9	3.7±1.7

Data from Corder et al.[3]

tase that had been purified from rat seminal vesicles, Ludvigson and Sorenson localized aldose reductase immunohistochemically in several areas of the kidney.[7] The most intense staining was seen in the inner medulla, whereas the outer medulla and the cortex stained faintly and inconsistently. Within the cortex, however, aldose reductase staining was found in the convoluted portions of the distal tubule and in the glomerular epithelial cells.

More recently, actual measurement of the polyol content has conclusively demonstrated that sorbitol accumulates in renal cortical tissue in the two traditional experimental models in which increased polyols would be anticipated. First, the level of galactitol is increased in renal cortex obtained from galactosemic rats, and second, the sorbitol content of glomeruli isolated from streptozotocin-diabetic rats is dramatically increased compared to that in glomeruli isolated from control animals.[8,9] The experiments in the latter report were performed with animals that had untreated diabetes for 6 to 9 weeks, were severely hyperglycemic, and had renal hypertrophy. Interestingly, the glomerular polyol content was less in 9-week than in 6-week animals—a finding compatible with cell leakage as a consequence of chronic polyol accumulation (Figure 5-1).[10] Since the diabetic-induced increase in glomerular sorbitol and the galactose-induced increase in renal cortical galactitol were both prevented by oral administration of an aldose reductase inhibitor, there is little doubt that the polyol pathway is operative in the renal cortex, and that it has the capacity to expand greatly in the presence of hyperglycemia or galactosemia. It is presumed that, in the glomerulus, this activity largely resides in the epithelial cells (podocytes) since these are the only glomerular cells that exhibit immunohistochemical reactivity to aldose reductase antibody.[7] Additionally, monolayer cultures of monkey kidney epithelium cells accumulated sorbitol when incubated in a high-glucose medium.[11] In this context, it is interesting that enlargement of renal cortical podocytes, suggestive of intracellular edema, and abnormal cytoplasmic extensions of the epithelial cell membranes have been described in diabetic glomerulosclerosis, and that the proteinuria occurring in diabetes has been related to morphologic alterations present in the epithelial cell foot processes.[12-14] Epithelial cells participate in basement membrane synthesis, and it is possible that enhanced polyol pathway activity disturbs metabolic processes

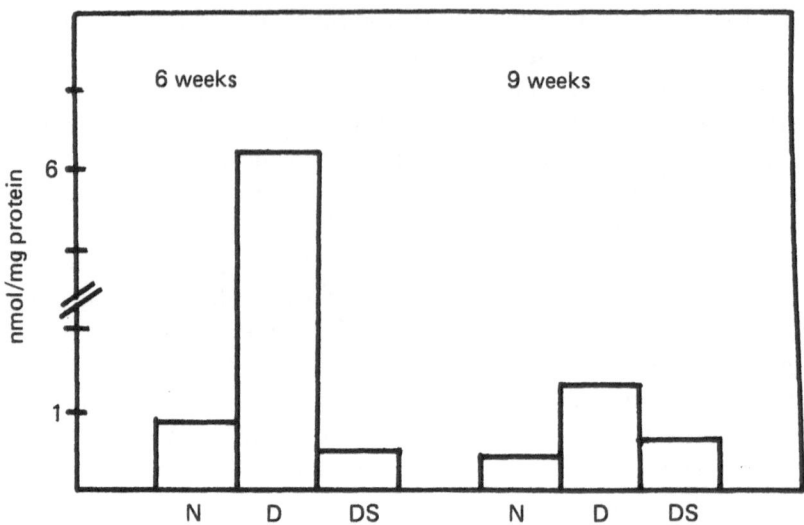

FIGURE 5-1   Polyol content of glomeruli isolated from control (N), diabetic (D) and Sorbinil-treated diabetic (DS) rats after 6 and 9 weeks of diabetes. Data from Beyer-Mears and Cohen.[9]

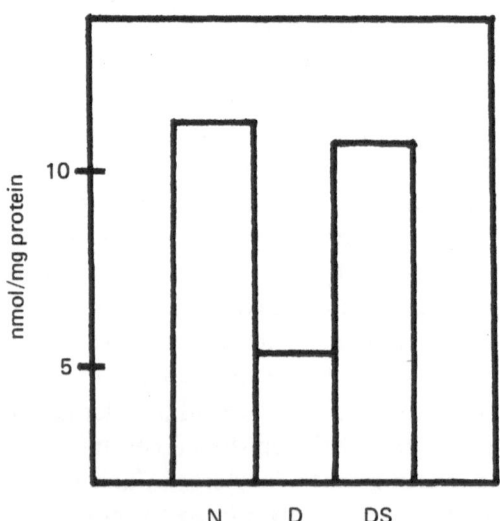

FIGURE 5-2   *Myo*-inositol content of glomeruli isolated from control (N), diabetic (D) and Sorbinil-treated diabetic (DS) rats after 9 weeks of diabetes. Data from Beyer-Mears and Cohen.[9]

involved with basement membrane production, although direct experimental evidence to support such a postulate is currently lacking. That the renal hypertrophy associated with galactosemia is reportedly diminished with aldose reductase inhibition[15] again suggests that the polyol pathway contributes in some way to renal cellular expansion.

Depletion of glomerular *myo*-inositol accompanies the sorbitol increase in streptozotocin diabetes (Figure 5-2). This change, as has been demonstrated in peripheral nerve, is prevented by treatment with an aldose reductase inhibitor,[9,16] indicating that the accumulation of polyol pathway intermediates in some way contributes to the fall in tissue *myo*-inositol content. The reduction in *myo*-inositol has been demonstrated by gas-liquid chromatographic analysis of homogenized, deproteinized isolated glomeruli, and by comparative in vitro radiolabeling experiments with [³H]-labeled *myo*-inositol of glomeruli isolated from control and streptozotocin-diabetic animals.[9,17] Incorporation of [³H]*myo*-inositol was significantly greater in glomeruli from diabetic rats compared to that in glomeruli from control animals at all incubational time periods examined (Figure 5-3). This increased radiolabeled *myo*-inositol incorporation in diabetic samples is compatible with a decrease in the amount of unlabeled (cold) *myo*-inositol in the cell, since the specific activity of the radiolabel would be increased when taken up by diabetic glomeruli in which there is depletion of (unlabeled) *myo*-inositol. A sodium-dependent, glucose-inhibited *myo*-inositol transport system has been demonstrated in isolated rat renal glomeruli, analogous to that identified in peripheral nerve.[18] Thus, reduced glomerular *myo*-inositol content in diabetes may reflect inhibition of *myo*-inositol uptake imposed by hyperglycemia, as has been proposed to explain similar findings in peripheral nerve. Glucose has also been shown to impair *myo*-inositol transport in isolated adipocytes and, presumably as a consequence, the formation of phosphatidylinositol in this tissue.[19]

Although the direct or indirect consequences of reduced *myo*-insotol and/or increased sorbitol on glomerular metabolism are not clear, the similarity of findings with respect to polyol and *myo*-inositol levels in glomeruli and peripheral nerve of animals with experimental diabetes suggested that other metabolic abnormalities might be common to both tissues. Indeed, examination of glomeru-

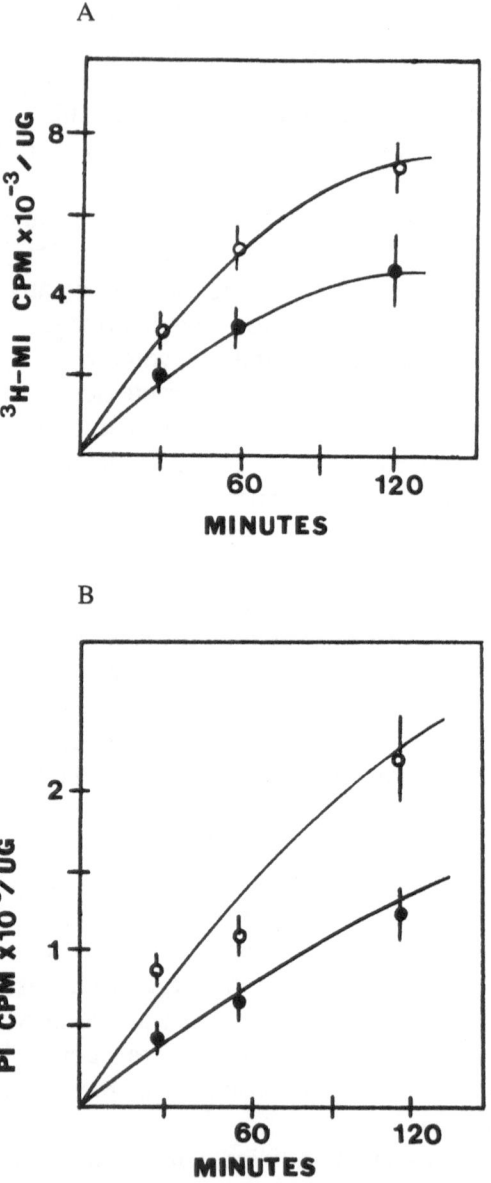

FIGURE 5-3 [³H]-*myo*-inositol (³H-MI) incorporation into glomerular phospholipids (A) and phosphatidylinositol (B) in control (●) and diabetic (○) samples. Reprinted with permission from Cohen MP et al: Effect of diabetes and sorbinil treatment on phospholipid metabolism in rat glomeruli. *Biochim Biophys Acta* 1986; 876:226–231.

lar sodium/potassium adenosine triphosphatase (Na/K-ATPase) revealed that activity was reduced in streptozotocin diabetes.[20] Again, this change was prevented by administration of an aldose reductase inhibitor, which has also been shown to correct decreased Na/K-ATPase activity in peripheral nerve in experimental diabetes.[21] It was also partially ameliorated by treatment with insulin, indicating that the decreased glomerular Na/K-ATPase activity derived from the abnormal metabolic milieu of diabetes rather than from a nephrotoxic effect of streptozotocin (Table 5-2). Although another communication reported that whole-kidney Na/K-ATPase was increased, rather than decreased, in streptozotocin-diabetic rats,[22] it appears that technical differences, choice of tissue sample, and duration of diabetes account for this dichotomy. For example, with increasing duration of untreated diabetes, superimposed hemodynamic and other factors may alter renal ATPase activity independently of the diabetic state. A rise in whole-kidney Na/K-ATPase activity may reflect a contribution from tubular components involved with compensatory hyperplasia and hypertrophy accompanying renal hyperfunction in diabetes.[23,24] Tubules and medulla contain large amounts of ATPase activity,[22,25] alterations in which could obscure more subtle changes in glomerular segments in response to specific conditions.

That inhibition of aldose reductase activity prevents both the fall in *myo*-inositiol levels and the reduction in Na/K-ATPase activity observed in glomeruli from diabetic animals suggests that these changes are linked not only to enhanced polyol pathway activity but also to each other. Among the proposed explanations for these findings is the postulate that the polyol pathway, or sorbitol accumulation, promotes *myo*-inositol efflux from the cell,[16] and that the

TABLE 5-2 Glomerular Na/K-ATPase Activity

Experimental Group	Activity (μmol Pi/mg/min)
Control	0.623 ± 0.054
Diabetic	0.425 ± 0.053
Sorbinil-treated diabetic	0.677 ± 0.107
Insulin-treated diabetic	0.542 ± 0.068

Data from Cohen et al.[20]

resulting reduction in *myo*-inositol limits the availability of this hex-itol for important inositol-containing phospholipid cycles associated with cell membrane structures and enzyme complexes. In this con-struct, reduced Na/K-ATPase activity might be related to decreased phosphatidylinositol, a membrane phospholipid that is intimately associated with this enzyme.[26] However, actual quantification of phosphatidylinositol in crude glomerular membranes revealed no significant differences in preparations from diabetic rats compared to concentrations in tissue from control animals (Table 5-3).[17] Fur-thermore, only about 25% of the phosphatidylinositol in renal microsomes appears necessary for activation of Na/K-ATPase.[26] This combination of findings makes it unlikely that changes in phos-phatidylinositol content, even if such were demonstrable, could account for the reduced glomerular Na/K-ATPase activity in acute experimental diabetes. On the other hand, phosphatidylinositol is a minor component, comprising less than 5% of total phospholipids in the kidney,[27] and it is both a precursor and a breakdown product of a cyclic series of reactions within the polyphosphoinositide response system.[28,29] Thus, modest changes in phosphatidylinositol, espe-cially if confined to the "hormone-sensitive pool," might escape detection by quantitative analysis.

Another interesting postulate that could help explain the preven-tion of *myo*-inositol depletion, and perhaps the decrease in Na/K-ATPase activity, by inhibition of aldose reductase activity relates to the finding that at least one aldose reductase inhibitor can interact directly with cell membranes. Binding of [³H]Sorbinil to crude membranes prepared from isolated rat glomeruli was found to be dose dependent, saturable, and inhibited by increasing concentra-tions of the unlabeled compound,[30] suggesting that the drug may

TABLE 5-3 Glomerular Phosphatidylinositol Content in Experimental Diabetes

Experimental Group	Phosphatidylinositol (μg Pi/mg dry wt)
Control	0.16±0.02
Diabetic	0.17±0.03
Sorbinil-treated diabetic	0.16±0.02

Data from Cohen et al.[17]

affect membrane-associated processes independent of, or in addition to, its aldose reductase-inhibiting properties.

A number of questions remain unresolved concerning the specific experimental findings cited above. For example, if one postulates that *myo*-inositol falls because it competes with the glucose transporter for cell entry,[31] how is the reduction in *myo*-inositol prevented by inhibition of aldose reductase in the face of persistent hyperglycemia? If decreased Na/K-ATPase activity relates to *myo*-inositol depletion, how does prevention of polyol accumulation by means of aldose reductase inhibition prevent the decrease in enzyme activity? More importantly, what, if any, is the role of the polyol pathway in the pathogenesis of diabetic nephropathy?

Among the biochemical changes believed to be pathogenetically linked to diabetic nephropathy are increased basement membrane collagen synthesis,[32-37] decreased basement membrane collagen turnover,[38-41] quantitative and/or qualitative abnormalities in basement membrane proteoglycans and particularly the glycosaminoglycan components,[42-49] and excess nonenzymatic glycosylation of basement membrane proteins.[50-55] These processes participate in the pathophysiology and structure/function changes manifest in the diabetic glomerulus as an accumulation of glomerular basement membrane in the peripheral capillary loops, an expansion of the mesangial matrix, a diminution in the anionic sites which regulate charge-selective permeability properties of the glomerular filtration barrier, and a disturbance in the organization, assembly, or interaction of macromolecules in the extracellular matrix. It remains to be established whether changes in the glomerular polyol and *myo*-inositol contents and in Na/K-ATPase activity influence basement membrane collagen and/or proteoglycan synthesis, or matrix organization, and if so, by what mechanisms.

The results of one early series of experiments offered promise toward uncovering a link between basement membrane synthesis and activation of the polyol pathway in diabetes. These experiments examined the effect of a quinoline derivative with aldose reductase-inhibiting properties on basement membrane collagen synthesis by the parietal yolk sac model system and by kidney glomeruli incubated in vitro.[56] This compound, GPA 1734, potently inhibited the formation of hydroxyproline and hydroxylysine, which reflects collagen production, in these tissues without having significant effect on total protein synthesis. The biochemical effect appeared to be at the level of the hydroxylating enzymes, since the compound inhibited the activity of prolyl hydroxylase partially purified from rat skin. Whether this effect is unique to that particular compound, and whether it is related to or independent of its aldose reductase-inhibiting property, are questions worth exploring.

Another recent study, reported in abstract form, examined the influence of treatment with Sorbinil on the gel electrophoretic patterns of urinary proteins in streptozotocin-diabetic rats.[57] Untreated diabetic rats excreted several proteins with molecular weights greater than albumin, in contrast to control animals, which excreted only albumin and low-molecular-weight proteins. Sorbinil treatment was reported to halt the progression of proteinuria and to restore the urinary protein pattern to normal. If these results are confirmed, the possibility that activation of the polyol pathway, or inhibition of aldose reductase, in glomerular epithelial cells affects the glomerular filtration barrier should certainly be examined. Similarly, the recent report that treatment with an aldose reductase inhibitor or dietary *myo*-inositol supplementation significantly reduced the increased glomerular filtration rate that is characteristically found in early, untreated experimental diabetes raises new questions concerning the relationships between glomerular polyol, *myo*-inositol, and hyperfiltration as well as their relative contributions to nephropathic processes.[58]

References

1. Hutton JC, Schofield PJ, Williams JF, et al: The localization of sorbitol pathway activity in the rat renal cortex and its relationship to the pathogenesis of the renal complications of diabetes mellitus. *Aust J Exp Biol Med Sci* 1975; 53:49–57.

2. Corder CN, Collins JG, Brannon TS, et al: Aldose reductase and sorbitol dehydrogenase distribution in rat kidney. *J Histochem Cytochem* 1977; 25:1–8.

3. Corder CN, Braughler JM, Culp PA: Quantitative histochemistry of the sorbitol pathway in glomeruli and small arteries of human diabetic kidney. *Folia Histochem Cytochem (Krakow)* 1979; 17:137–146.

4. Kern TS, Engerman RL: Immunohistochemical distribution of aldose reductase. *Histochem J* 1982; 14:507–515.

5. Gabbay KH, O'Sullivan JB: The sorbitol pathway in diabetes and galactosemia: Enzyme and substrate localization and changes in the kidney. *Diabetes* 1968; 17:300A.

6. Gabbay KH: The sorbitol pathway and complications of diabetes. *N Engl J Med* 1973; 288:831–836.

7. Ludvigson MA, Sorenson RL: Immunohistochemical localization of aldose reductase. *Diabetes* 1980; 29:450–459.

8. Beyer-Mears A, Nicolas-Alexandre J, Cruz E: Sorbinil inhibition of renal aldose reductase. *Fed Proc* 1983; 42:858.

9. Beyer-Mears A, Ku L, Cohen MP: Glomerular polyol accumulation in diabetes and its prevention by oral sorbinil. *Diabetes* 1984; 33:604–607.

10. Reddy VN, Schauss D, Chakrapani B, et al: Biochemical changes associated with the development and reversal of cataracts. *Exp Eye Res* 1976; 23: 483–493.

11. Hutton JC, Williams JF, Schofield PJ, et al: Polyol metabolism in monkey-kidney epithelial cell cultures. *Eur J Biochem* 1974; 49:347–353.

12. Cohen AH, Mampaso F, Zamboui L: Glomerular podocyte degeneration in human renal disease. *Lab Invest* 1977; 37:40–42.

13. Jones DB: Correlative scanning and transmission electron microscopy of glomeruli. *Lab Invest* 1977; 37:569–578.

14. Jones DB: SEM of human and experimental renal disease. *Scan Electron Microsc* 1979; 3:679–689.

15. Beyer-Mears A, Cruz E, Dillon P, et al: Diabetic renal hypertrophy diminished by aldose reductase inhibition. *Fed Proc* 1983; 42:505.

16. Finegold D, Lattimer SA, Nolle S, et al: Polyol pathway activity and *myo*-inositol metabolism. A suggested relationship in the pathogenesis of diabetic neuropathy. *Diabetes* 1983; 32:988–992.

17. Cohen MP, Klepser H, Cua E: Effect of diabetes and Sorbinil treatment on phospholipid metabolism in rat glomeruli. *Biochim Biophys Acta* 1986; 876:226–231.

18. Ulbrecht JS, Bensen JT, Greene DA: Sodium-dependent *myo*-inositol uptake in isolated renal glomeruli: Possible inhibition by glucose. *Diabetes* 1985; 34(suppl 1):13A.

19. Kaplan SA, Lee W-NP, Scott ML: Glucose inhibits *myo*-inositol transport and phosphatidylinositol formation in adipocytes. *Diabetes* 1985; 34(suppl 1): 183A.

20. Cohen MP, Dasmahapatra A, Shapiro E: Reduced glomerular sodium/potassium adenosine triphosphatase activity in acute streptozotocin diabetes. *Diabetes* 1985; 34:1071–1074.

21. Greene DA, Lattimer SA: Action of Sorbinil in diabetic peripheral nerve: Relationship of polyol (sorbitol) pathway inhibition to a *myo*-inositol-mediated defect in sodium-potassium ATPase activity. *Diabetes* 1984; 33:712–716.

22. Finegold DN, Nolle SS, Lattimer S, et al: Alterations in renal ouabain sensitive Na/K-ATPase activity in streptozotocin diabetes. *Clin Res* 1984; 32:395A.

23. Ku DD, Meezan E: Increased renal tubular sodium pump and Na, K-adenosine triphosphatase in streptozotocin-diabetic rats. *J Pharmacol Exp Ther* 1984; 229:664–670.

24. Seyer-Hansen K: Renal hypertrophy in experimental diabetes: Relation to severity of diabetes. *Diabetologia* 1977; 13:141-143.

25. Lo C-S, August TR, Liberman UA, et al: Dependence of renal (Na.K-ATPase)-adenosine triphosphatase activity on thyroid status. *J Biol Chem* 1976; 251:7826-7833.

26. Roelofsen B, Trip MVL-S: The fraction of phospatidylinositol that activates the (Na$^+$/K$^+$)-ATPase in rabbit kidney microsomes is closely associated with the enzyme protein. *Biochim Biophys Acta* 1981; 647:302-306.

27. Toback FG: Phosphatidylcholine metabolism during renal growth and regeneration. *Am J Physiol* 1984; 246:F249-F259.

28. Farese RV: Phosphoinositide metabolism and hormone action. *Endocr Rev* 1983; 4:78-95.

29. Nishizuka Y: Turnover of inositol phospholipids and signal transduction. *Science* 1984; 225:1365-1370.

30. Cohen MP, Klepser H: Binding of an aldose reductase inhibitor to renal glomeruli. *Biochem Biophys Res Commun* 1985; 129:530-535.

31. Greene DA, Lattimer SA: Sodium and energy-dependent uptake of *myo*-inositol by rabbit peripheral nerve: Competitive inhibition by glucose and lack of insulin effect. *J Clin Invest* 1982; 70:1009-1018.

32. Cohen MP, Khalifa A: Renal glomerular collagen synthesis in streptozotocin diabetes. Reversal of increased basement membrane synthesis with insulin therapy. *Biochim Biophys Acta* 1977; 500:395-404.

33. Cohen MP: Glomerular basement membrane synthesis in streptozotocin diabetes, in Podolsky S, Viswanatha M (eds): *Secondary Diabetes, the Spectrum of the Diabetic Syndromes*, pp 541-551. New York, Raven Press, 1980.

34. Cohen MP, Dasmahapatra A, Wu VY: Deposition of basement membrane in vitro by normal and diabetic renal glomeruli. *Nephron* 1981; 27:146-151.

35. Khalifa A, Cohen MP: Glomerular protocollagen lysyl hydroxylase activity in streptozotocin diabetes. *Biochim Biophys Acta* 1975; 386:332-339.

36. Ristelli J, Koivisto VA, Akerblom HK, et al: Intracellular enzymes of collagen biosynthesis in rat kidney with streptozotocin diabetes. *Diabetes* 1976; 25:1066-1070.

37. Spiro RG, Spiro MJ: Effect of diabetes on the biosynthesis of renal glomerular basement membrane. Studies on the glucosyltransferase. *Diabetes* 1971; 20:641-648.

38. Brownlee M, Spiro RG: Glomerular basement membrane metabolism in the diabetic rat. *In vivo* studies. *Diabetes* 1979; 28:121-125.

39. Cohen MP, Surma ML, Wu VY: In vivo biosynthesis and turnover of glomerular basement membrane in diabetic rats. *Am J Physiol* 1982; 242:F385-F389.

40. Romen W, Heck T, Rauscher G, et al: Glomerular basement membrane turnover in young, old, and streptozotocin-diabetic rats. *Renal Physiol* 1980; 3:324-329.

41. Romen W, Lange H-W, Hempel K, et al: Studies on collagen metabolism in rat. II. Turnover and amino acid composition of the collagen of glomerular basement membrane in diabetes mellitus. *Virchows Arch [Cell Pathol]* 1981; 36:313–320.

42. Martines-Hernandez A, Amenta P: The basement membrane in pathology. *Lab Invest* 1983; 48:656–677.

43. Cohen MP, Surma ML: [^{35}S]-Sulfate incorporation into glomerular basement membrane glycosaminoglycans is decreased in experimental diabetes. *J Lab Clin Med* 1981; 98:715–722.

44. Cohen MP, Surma ML: Effect of diabetes on in vivo metabolism of [^{35}S]-labeled glomerular basement membrane. *Diabetes* 1984; 33:8–12.

45. Kanwar YS, Rosenzweig LJ, Linker A, et al: Decreased de novo synthesis of glomerular proteoglycans in diabetes: Biochemical and autoradiographic evidence. *Proc Natl Acad Sci USA* 1983; 80:2272–2275.

46. Rohrbach DH, Wagner CW, Star VL, et al: Reduced synthesis of basement membrane heparan sulfate proteoglycan in streptozotocin-induced diabetic mice. *J Biol Chem* 1983; 258:11672–11677.

47. Parathasarathy N, Spiro RG: Effect of diabetes on the glycosaminoglycan component of the human glomerular basement membrane. *Diabetes* 1982; 31:738–741.

48. Rohrbach DH, Hassell JR, Kleinman HK, et al: Alterations in the basement membrane (heparan sulfate) proteoglycan in diabetic mice. *Diabetes* 1982; 31:185–188.

49. Wu VY, Cohen MP: Platelet factor 4 binding to glomerular microvascular matrix. *Biochim Biophys Acta* 1984; 797:76–82.

50. Vogt BW, Schleicher ED, Wieland OH: ε-Amino-lysine bound glucose in human tissues obtained at autopsy: Increase in diabetes mellitus. *Diabetes* 1983; 31:1123–1127.

51. Cohen MP, Urdanivia E, Surma M, et al: Increased glycosylation of glomerular basement membrane collagen in diabetes. *Biochem Biophys Res Commun* 1980; 95:765–769.

52. Cohen MP, Wu VY: Identification of specific amino acids in diabetic glomerular basement membrane subject to nonenzymatic glycosylation in vivo. *Biochem Biophys Res Commun* 1984; 100:1549–1554.

53. Perejda AJ, Uitto J: Nonenzymatic glycosylation of collagen and other proteins: Relationship to the development of diabetic complications. *Coll Rel Res* 1982; 2:81–88.

54. Trueb B, Fluckiger R, Winterhalter KH: Nonenzymatic glycosylation of basement membrane collagen in diabetes mellitus. *Coll Rel Res* 1984; 4:239–251.

55. Uitto J, Perejda AJ, Grant GA, et al: Glucosylation of human glomerular basement membrane collagen: Increased content of hexose in ketoamine linkage and unaltered hydroxylysine-o-glycosides in patients with diabetes. *Connect Tiss Res* 1982; 10:287–296.

56. Maragoudakis ME, Kelinsky H, Wasvary J, et al: Inhibition of basement membrane synthesis and aldose reductase activity by GPA 1734. *Fed Proc* 1976; 35:679.

57. Beyer-Mears A, Varagiannis E, Cruz E: Effect of Sorbinil on reversal of proteinuria. *Diabetes* 1985; 35(suppl 1):101A.

58. Goldfarb S, Simmons DA, Kern E: Amelioration of glomerular hyperfiltration in acute experimental diabetes by dietary *myo*-inositol and by an aldose reductase inhibitor. *Clin Res* 1986; 34:725A.

CHAPTER 6

Aldose Reductase and the Vascular System

Microvasculature

Using a model that they developed to study the influence of various manipulations on vascular permeability, Williams and co-workers have described some interesting effects of experimental diabetes, galactose feeding, and the aldose reductase inhibitor Sorbinil. This model uses new vessels formed by angiogenesis in granulation tissue induced by subcutaneous implantation of sterile polyester fabric, and measures the permeation of ^{125}I-labeled albumin in this tissue by comparing the ratios of radioactive albumin to ^{51}Cr-labeled erythrocytes in the tissue and in the blood at a fixed time after injection of both radioisotopes.[1-3] Vascular permeation of albumin in granulation tissue from female BB/W rats and from male streptozotocin-diabetic rats was significantly greater than that in granulation tissue induced by the same procedure in control animals.[2-4] Since such differences in albumin permeation were not observed in other tissues, such as skin, fat, or aorta, taken from control and diabetic rats, the investigators proposed that the functional integrity of the vasculature is more likely to be impaired in recently formed than in older vessels. Galactose feeding also produced increased ^{125}I-labeled albumin permeation in these angiogenic new vessels and in eyes, even though it had previously been found that albumin permeation was not increased in the eyes of diabetic rats.[5]

The finding that both diabetes and galactose feeding caused increased vascular permeability in this model suggested that polyol accumulation might be involved in the observed defect in albumin permeation in the newly formed vessels. In subsequent experiments, male diabetic rats were treated with Sorbinil, commencing the day of implantation, or were castrated 10 days before the polyester fabric was implanted.[6,7] Either manipulation was successful in preventing the diabetes-induced increase in vascular permeability, indicating that aldose reductase participated in the development of this abnormality and suggesting that sex steroids influence the activity or expression of this enzyme. Castration also markedly lessened the diabetes-associated increase in sorbitol levels and decrease in *myo*-inositol content in granulation tissue. It was further noted that granulation tissue from diabetic animals had increased collagen cross-linking, assessed by percent solubility in 0.5 M acetic acid, and that this change was prevented by castration of diabetic animals but was unaffected by Sorbinil treatment.[8]

Macrovasculature

Thoracic aorta produces increased amounts of sorbitol and fructose when incubated in media containing high glucose concentrations.[9,10] There is a progressive increase in tissue sorbitol content when segments of thoracic aorta are exposed to increasing concentrations of glucose (0 to 50 mM). Notably, aortic sorbitol content also increased when incubations were conducted in the presence of epinephrine (2 μg/mL), isoproterenol (4 μg/mL), dibutyryl-3′,5′-adenosine monophosphate (1×10^{-4} M), ouabain (1×10^{-5} M), or angiotensin II (1 μg/mL), although the mechanism by which these agents promote sorbitol accumulation is obscure. In analogy to the lens, it was initially believed that osmotic and ionic changes resulting from polyol accumulation and leading to fibrosis could help explain the high incidence of atherosclerosis in the diabetic population. Water content is, in fact, increased when rabbit thoracic aorta smooth muscle cells or aortic segments are incubated for 2 hours with a 20 to 50 mM glucose concentration, and this increased water content occurs without a significant increase in the tissue inulin space. Osmotic-induced tissue damage could occur, and the explanation may thus still pertain, at least in part. Glucose-induced polyol accumulation in aortic smooth muscle cells is associated with a reduction in the *myo*-inositol content, and inhibition of aldose reductase (with ibuprofen) restores *myo*-inositol levels even when glucose concentration in the incubational media is kept at 50 mM.[11] Thus, as in peripheral nerve and the renal glomerulus, changes resulting from *myo*-inositol depletion may exert a deleterious influence on aortic wall metabolism in diabetes, and the relationship between the polyol pathway and the development of macrovascular complications is probably as complex as it is in other tissues.

In this context, the metabolic changes that occur when the aorta is deprived of its normal extracellular *myo*-inositol concentrations are of interest.[12] When normal rabbit aortic intima-media preparations are incubated without *myo*-inositol, oxygen uptake falls concomitant with depletion of endogenous *myo*-inositol. Comparison of [^{14}C]glycerol incorporation into aortic phosphatidylinositol in the presence versus the absence of media *myo*-inositol unmasked a restricted pool of tissue phosphatidylinositol that has rapid turnover, reflecting basal phosphatidylinositol hydrolysis. Thus, maintenance of normal plasma levels of *myo*-inositol is required to prevent inhibition of a discrete component of basal de novo phosphatidylinositol synthesis. Additionally, normal levels of *myo*-inositol are required for maintenance of a specific component of resting energy utilization by aortic tissue. This energy utilization largely derives from oxygen consumption resulting from Na/K-ATPase activity, and the resting Na/K-ATPase activity falls when the tissue is depleted of *myo*-inositol. These results link tissue *myo*-inositol content, phosphatidylinositol metabolism, and Na/K-ATPase activity, and provide insight into the relationship between these metabolic changes that has not been readily afforded by examination of the effect of diabetes on phosphatidylinositol content or metabolism in other tissues (see Chapters 4 and 5).

Other metabolic effects associated with increased flux through the polyol pathway in the aortic wall as a result of increased intracellular glucose concentration include a reduced oxygen uptake, enhanced glycolysis, and an increase in the ratio of the concentrations of lactate to pyruvate.[10] These changes are similar to those observed in human erythrocytes exposed to high glucose concentration in vitro, and may reflect alterations in the redox states of the diphosphopyridine nucleotide cofactors.[13]

Erythrocytes

The intrinsic capability of normal red blood cells to undergo shape adaptation permits them to traverse capillaries with diameters smaller than their own. This property of erythrocyte deformability depends on compositional and elastic features of the membrane, and on the composition of intracellular and plasmatic fluids. The deformability of red cells from patients with diabetes mellitus is impaired, and alterations in the filtration properties of erythrocytes have been consistently demonstrated in diabetes.[14-22] In addition to other rheologic abnormalities associated with diabetes, such as increased plasma viscosity,[18,23] decreased membrane fluidity,[19] and exaggerated erythrocyte aggregability and adhesiveness,[17,24,25] decreased erythrocyte deformability has been implicated in the pathogenesis of diabetic microvascular complications. For example, one study found that deformability was significantly less in diabetic patients with widespread complications than in those with minimal or no complications.[23] A variety of factors, including ATP or calcium levels, plasma pH, and intracellular or plasma osmolarity, can influence deformability, and changes in this red cell property that occur in diabetes have been related to abnormal membrane fluidity,[16] altered composition or saturation of membrane lipids,[26-28] and intraerythrocytic aberrations due to sorbitol accumulation[29] or increased glycosylated hemoglobin content.[14,18] In fact, deformability has been reported to be reduced in proportion to the degree of intraerythrocytic sorbitol accumulation (Figure 6-1).[29]

Erythrocytes from patients with diabetes contain significantly higher levels of sorbitol than do cells from nondiabetic subjects.[30-32] The activity of aldose reductase, measured as NADPH-oxidizing activity, is reportedly higher in red cells from diabetic patients with retinopathy or cataracts than it is in erythrocytes from suitable control patients, and shows a positive correlation with Hb A_{1c} levels in

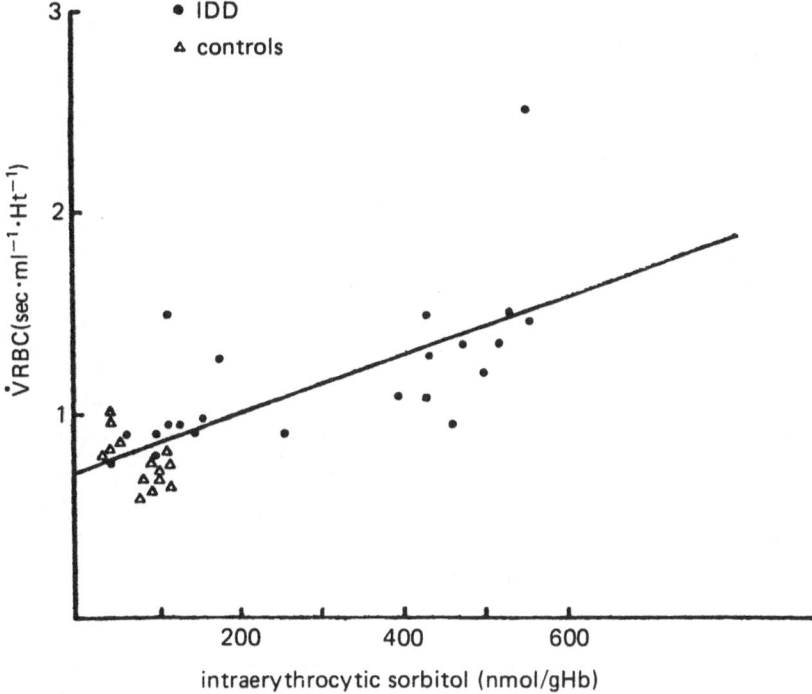

FIGURE 6-1   Correlation between red cell sorbitol and filtration index (VRBC) in control (△) and diabetic (●) subjects. Reprinted with permission from Caradente O et al: Role of red cell sorbitol as determinant of reduced erythrocyte filtrability in insulin defendant diabetes. *Acta Diabetol Lat* 1982; 19(4):359–369.

diabetic cataract patients.[33-35] Similarly, activity of the aldose reductase-like glyceraldehyde dehydrogenase is elevated in diabetic patients with retinopathy and cataract.[36] Sorbitol is also produced in normal red cells when they are incubated in a glucose-containing medium, and there is some evidence that red cells from diabetic subjects accumulate more sorbitol than do cells from normal individuals exposed in vitro to the same incubation system.[13,30,31,37] Inhibition of aldose reductase activity prevents erythrocyte accumulation of sorbitol in vitro,[38] and treatment of diabetic patients with an aldose reductase inhibitor normalizes the red cell sorbitol content.[32,39-42] However, inhibition of aldose reductase does not affect the concentration of 2,3-diphosphoglycerate (2,3-DPG) in erythrocytes or the oxygen affinity of hemoglobin.[40] This is of interest in view of the suggestion that increased polyol pathway activity decreases the ratio of NAD to NADH, thereby decreasing 2,3-DPG levels and, in turn, influencing oxyhemoglobin dissociation.[13,43] In any event, a relationship between 2,3-DPG concentration and red cell deformability has not been established. It remains to be determined whether there are other potentially deleterious metabolic effects within the erythrocyte that derive from changes in the oxidized and reduced forms of the NAD^+ and $NADP^+$ redox systems that may be initiated by enhanced polyol pathway activity.

Erythrocyte deformability is also significantly diminished in experimental diabetes.[44] Of considerable interest is the finding that oral administration of an aldose reductase inhibitor to rats with streptozotocin diabetes restores deformability, although not quite to normal levels[44,45] (Figure 6-2). This correction was seen despite comparable levels of hyperglycemia in untreated and Sorbinil-treated rats, indicating that the effect of the agent was not mediated through an influence on plasma viscosity. Further support for such

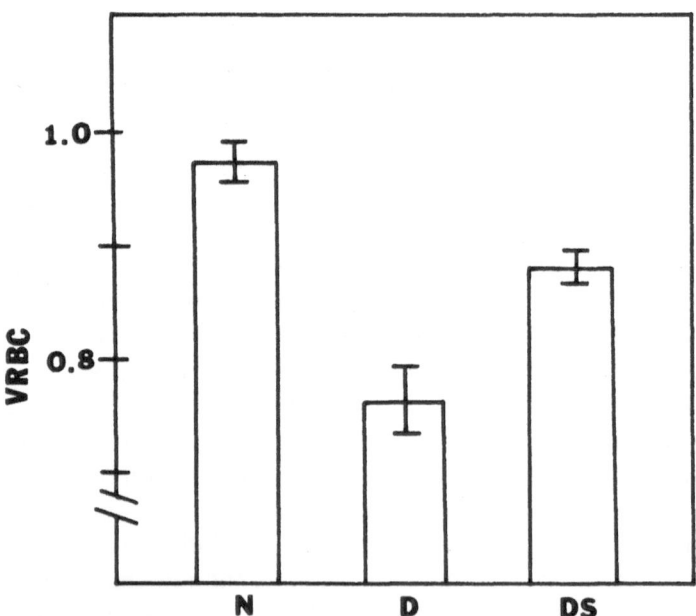

FIGURE 6-2   Erythrocyte deformability (VRBC) in whole blood from control (N), diabetic (D), and Sorbinil-treated diabetic (DS) rats. Data from Robey et al.[45]

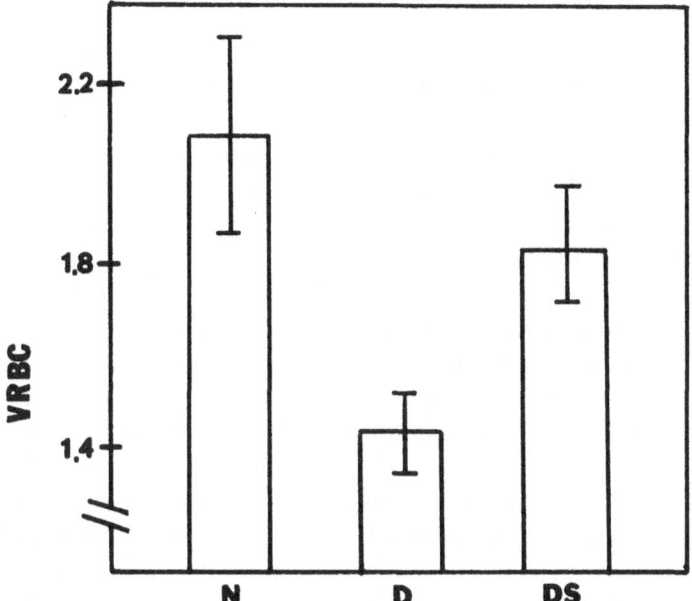

FIGURE 6-3   Deformability in washed erythrocytes from control (N), diabetic (D), and Sorbinil-treated diabetic (DS) rats. Data from Robey et al.[45]

an interpretation was offered with the finding that the defect in erythrocyte deformability in diabetes, and its correction with aldose reductase inhibition, persisted even when red cells were washed (to eliminate the hyperglycemic milieu and other plasma factors of potential influence) and resuspended before measuring their deformability (Figure 6-3). However, persistent hyperglycemia may explain the finding that erythrocyte deformability in whole blood remained less than that of control in red cells from Sorbinil-treated animals.

The mechanism by which the aldose reductase inhibitor Sorbinil prevents the decreased erythrocyte deformability associated with experimental diabetes is not clear. Sorbinil reduces the elevated red cell sorbitol levels in diabetes, and normalization of the intra-erythrocytic sorbitol content may in part explain the ability of Sorbinil to prevent reduced erythrocyte deformability. This explanation presumes that osmotic effects attendant to sorbitol accumulation contribute to the decreased deformability. However, there is some question whether absolute levels of sorbitol achieved in the red cell are sufficient to produce an osmotic effect. Intraerythrocytic sorbitol concentrations are in the micromolar range,[13,39] whereas those reported for peripheral nerve and lens are 10- to 100-fold higher, and it is therefore unlikely that polyol accumulation in the red cell is sufficient to be osmotically active. Further, the addition of Sorbinil in vitro to erythrocytes incubated with glucose (500 mg/dL) prevented the sorbitol accumulation but not the reduced deformability observed with exposure to high glucose concentration.[38] It appears that neither the reduced deformability associated with experimental diabetes nor its correction by aldose reductase inhibition can be fully explained by osmotic factors accompanying erythrocytic sorbitol accumulation, and other influences must be sought.

In contrast to findings in several other tissues (nerve, glomeruli, retinal vasculature), the elevated erythrocyte sorbitol content in diabetic samples is apparently not accompanied by a diminution in red cell myo-inositol levels, at least in human diabetes.[32] myo-Inositol levels in erythrocytes from streptozotocin-diabetic rats do not reflect changes in myo-inositol content in nerve or lens.[46] Although this finding suggests that myo-inositol depletion is not an obligatory concomitant of polyol accumulation in all cells, the reason for the relationship between these two changes in other tissues has not been clearly delineated.

References

1. Chang K, Uitto J, Rowold EA, et al: Increased collagen cross-linkages in experimental diabetes. *Diabetes* 1980; 29:778–781.

2. Williamson JR, Rowold E, Chang K, et al: Albumin permeation of new (granulation tissue) vessels is increased in diabetic rats. *Diabetes* 1984; 33(suppl 1): 3A.

3. Kilzer P, Chang K, Marvel J, et al: Albumin permeation of new vessels is increased in diabetic rats. *Diabetes* 1985; 34:333–336.

4. Williamson JR, Chang K, Rowold E, et al: Sorbinil prevents diabetes-induced increases in vascular permeability but does not alter collagen cross-linking. *Diabetes* 1985; 34:703–705.

5. Chang K, Rowold E, Marvel J, et al: Increased [125I]-albumin permeation of vessels in rats fed galactose. *Diabetes* 1985; 34(suppl 1):40A.

6. Williamson JR, Chang K, Rowold E, et al: Diabetes-induced increases in vascular permeability are prevented by castration and by Sorbinil. *Diabetes* 1985; 34(suppl 1):108A.

7. Williamson JR, Chang C, Rowold E, et al: Diabetes-induced increases in vascular permeability and changes in granulation tissue levels of sorbitol, *myo*-inositol, *chiro*-inositol, and *scyllo*-inositol are prevented by Sorbinil. *Metabolism* 1986; 35(suppl 1):41–45.

8. Williamson JR, Rowold E, Chang K, et al: Sex steroid dependency of diabetes-induced changes in polyol metabolism, vascular permeability, and collagen cross-linking. *Diabetes* 1986; 35:20–27.

9. Clements RS, Morrison AD, Winegrad AI: Polyol pathway in aorta. Regulation by hormones. *Science* 1969; 166:1007–1008.

10. Morrison AD, Clements RS Jr, Winegrad AI: Effects of elevated glucose concentrations on the metabolism of the aortic wall. *J Clin Invest* 1972; 51:3114–3123.

11. Morrison AD: Linkage of polyol pathway activity and *myo*-inositol in aortic smooth muscle. *Diabetes* 1985; 34(suppl 1):12A.

12. Simmons DA, Kern EFO, Winegrad AI, et al: Basal phosphatidylinositol turnover controls aortic Na$^+$/K$^+$ ATPase activity. *J Clin Invest* 1986; 77:503–513.

13. Travis SF, Morrison AD, Clements RS Jr, et al: Metabolic alterations in the human erythrocyte produced by increases in glucose concentration. *J Clin Invest* 1971; 50:2104–2112.

14. Cataliotti R: Spectroscopic evidence of structural modifications in erythrocyte membranes of diabetic patients. *Stud Biophys* 1978; 73:199.

15. Hoare EM, Barnes AJ, Dormandy JA: Abnormal blood viscosity in diabetes mellitus and retinopathy. *Biorheology* 1978; 13:21.

16. Kamada T, Otsuji S: Low-levels of erythrocyte membrane fluidity in diabetic patients. A spin label study. *Diabetes* 1983; 32:585–591.

17. Schmid-Schonbein H, Volger E: Red cell aggregation and red cell deformability in diabetes. *Diabetes* 1976; 25:897–902.

18. McMillan DE, Utterback NG, La Puma J: Reduced erythrocyte deformability in diabetes. *Diabetes* 1978; 27:895–901.

19. Baba Y, Kai M, Kamada T, et al: Higher levels of erythrocyte membrane microviscosity in diabetes. *Diabetes* 1979; 28:1138–1140.

20. Dormandy JA, Hoare E, Colley U, et al: Clinical, haemodynamic, rheological and biochemical findings in 126 patients with intermittent claudication. *Br Med J* 1973; IV:576–581.

21. Dormandy JA, Hoare E, Dhatab AH, et al: Prognostic significance of rheological and biochemical findings in patients with intermittent claudication. *Br Med J* 1973; IV:581–583.

22. Pozza G, Cordaro C, Carandente O, et al: Study on relationship between erythrocyte filtration and other risk factors in diabetic angiopathy. *Ric Clin Lab* 1981; 11(suppl 1):317–326.

23. Barnes AJ, Locke P, Scudder PR, et al: Is hyperviscosity a treatable component of diabetic micro-circulatory disease? *Lancet* 1971; 2:789–791.

24. Satoh M, Imaizumi K, Bessho T, et al: Increased erythrocyte aggregation in diabetes and its relationship to glycosylated hemoglobin and retinopathy. *Diabetologia* 1984; 27:517–521.

25. Wautier JL, Paton RC, Wautier M-P, et al: Increased adhesion of erythrocytes to endothelial cells in diabetes mellitus and its relation to vascular complications. *N Engl J Med* 1981; 305:237–242.

26. Cooper RA: Abnormalities of cell membrane fluidity in the pathogenesis of disease. *N Engl J Med* 1977; 297:371–377.

27. Eck MG, Wynn JO, Carter WJ, et al: Fatty acid desaturation in experimental diabetes mellitus. *Diabetes* 1979; 28:479–485.

28. Clark DL, Harnel FG, Queener SF: Changes in renal phospholipid fatty acids in diabetes mellitus; correlation with changes in adenylate cyclase activity. *Lipids* 1983; 18:696–705.

29. Carandente O, Colombo R, Girardi AM, et al: Role of red cell sorbitol as determinant of reduced erythrocyte filtrability in insulin dependent diabetics. *Acta Diabetol Lat* 1982; 19(4):359–369.

30. Malone JI, Knox G, Harvey C: Sorbitol accumulation is altered in Type I (insulin-dependent) diabetes mellitus. *Diabetologia* 1984; 27:509–513.

31. Malone J, Knox G, Benford S, et al: Red cell sorbitol—an indicator of diabetic control. *Diabetes* 1980; 29:861–864.

32. Popp-Snijders C, Lomecky-Janousek MZ, Schouten JA, et al: *Myo*-inositol and sorbitol in erythrocytes from diabetic patients before and after Sorbinil treatment. *Diabetologia* 1984; 27:514–516.

33. Crabbe MJC, Basak Halder A: Affinity chromatography of bovine lens aldose reductase, and a comparison of some kinetic properties of the enzyme from lens and human erythrocyte. *Biochem Soc Trans* 1980; 8:194–195.

34. Basak Halder A, Wolff S, Ting H-H, et al: An aldose reductase from the human erythrocyte. *Biochem Soc Trans* 1980; 8:644–645.

35. Crabbe MJC, Brown AJ, Peckar CO, et al: NADPH-oxidizing activity in lens and erythrocytes in diabetic and nondiabetic patients with cataract. *Br J Ophthalmol* 1983; 67:696–699.

36. Crabbe MJC, Halder AB, Peckar CO, et al: Erythrocyte glyceraldehyde-reductase levels in diabetes with retinopathy and cataracts. *Lancet* 1980; 2:1268–1270.

37. Morrison AD, Clements RS Jr, Travis SB, et al: Glucose utilization by the polyol pathway in human erythrocytes. *Biochem Biophys Res Commun* 1970; 40:199–205.

38. Robey C, Dasmahapatra A, Cohen MP: In vitro effects of hyperglycemia and Sorbinil on erythrocyte (RBC) deformability. *Diabetes* 1986; 35:108A.

39. Malone JI, Leavengood H, Peterson MJ, et al: Red blood cell sorbitol as an indicator of polyol pathway activity. *Diabetes* 1984; 33:45–49.

40. Martyn CN, Matthews DM, Popp-Snijders C, et al: Effects of sorbinil treatment on erythrocytes and platelets of persons with diabetes. *Diabetes Care* 1986; 9:36–39.

41. Puhakainen E, Saamanen AM, Lehtinen J, et al: The effect of aldose reductase inhibitor (sorbinil) on erythrocyte sorbitol concentration in diabetic neuropathy. *Acta Endocrinol* 1983; 103(suppl 257):56.

42. Raskin P, Rosenstock J, Challis P, et al: Effect of tolrestat on RBC sorbitol levels in diabetic subjects. *Diabetes* 1985; 34(suppl 1):7A.

43. Huehns ER: Disorders of carbohydrate metabolism in the red blood corpuscle. *Clin Endocrinol Metab* 1976; 5:651–674.

44. Robey C, Dasmahapatra A, Cohen MP, et al: Sorbinil prevents decreased erythrocyte deformability in diabetes. *Diabetes* 1985; 34:161A.

45. Robey C, Dasmahapatra A, Cohen MP, et al: Sorbinil prevents decreased erythrocyte deformability in diabetes mellitus. Manuscript submitted, 1986.

46. Stockard C, Clements R: Usefulness of blood elements in the prediction of the fructose and *myo*-inositol content of nerve and lens. *Diabetes* 1984; 33(suppl 1):89A.

CHAPTER 7

Clinical Trials

Clinical studies examining the effect of aldose reductase inhibitors on symptoms and signs of diabetic neuropathy in human patients have been conducted for the past several years. To date, there are nine full reports published in the standard medical literature describing the results of these studies.[1-9] Several large-scale clinical trials, sponsored by pharmaceutical houses, to evaluate the safety and efficacy of their respective agents are either commencing, ongoing, nearing completion, or currently undergoing data analysis. However, aside from interim presentations, abstracts, and comments offered at various professional meetings or at special conferences sponsored by the involved pharmaceutical concern,[10-16] the coordinated findings of such multicenter, cooperative clinical trials are not yet available for review.

The earliest studies reported experience with alrestatin, an isoquinoline derivative developed by Ayerst Laboratories. Although this drug proved hepatotoxic, and Ayerst has replaced it with the less toxic tolrestat, it appeared to alleviate neuropathic symptoms in many of the 37 patients in whom it was used for which there is published information.[1-4] The first study gave alrestatin, 50 mg/kg of body weight per day IV in four divided doses to two patients with non-insulin-dependent diabetes, and 1 g four times per day orally for 30 days to four patients with adult-onset diabetes and severe peripheral neuropathy of 5 months' to 6 years' duration.[1] The patients receiving intravenous therapy reported marked improvement in clinical symptoms 2 days following the start of the infusions, with subjective improvement lasting up to 3 weeks after the infusions were discontinued. However, there were no significant changes in peripheral nerve conduction velocities and no objective improvements on neurologic examination in these patients. The four patients who received oral alrestatin experienced no beneficial symptomatic effect and had no objective improvements on neurologic examination or in peripheral nerve conduction velocities.

Subsequently, 10 patients between the ages of 43 and 67 years who had had diabetes for a mean duration of 14.4 years were given alrestatin, 50 mg/kg of body weight IV in four divided doses, daily for 5 days.[2] All of the patients had clinical manifestations of diabetic neuropathy, including paresthesias and muscle fatigability; seven claimed that symptoms were improved by the end of the treatment period, although there was no placebo control group. Motor nerve conduction velocities after treatment did not differ from baseline values, but posttreatment sensory conduction velocities increased substantially in four patients. The drug had no adverse effect on blood or urine chemistries.

Another study evaluated the effect of alrestatin, up to 8 g/day in divided oral doses for 12 weeks, on clinical and electrophysiologic parameters of nerve function in 14 patients, 6 of whom had insulin-dependent diabetes and 8 of whom had non-insulin-dependent diabetes.[3] The mean age of all patients receiving alrestatin was 55 years, and the mean duration of diabetes was 18.6 years. A second group of 16 patients, similarly composed and with comparable mean age and duration of diabetes, received placebo tablets for a 12-week period, and the study was conducted in a randomized, double-blind manner. Variables used to assess response included quantitative measurements of sensory thresholds for vibratory, tactile, and thermal stimuli; appreciation of warm-cold difference with a thermostimulator; motor and sensory conduction velocities; and galvanic skin response to a startle reaction. Of the 14 patients on alrestatin and the 14 patients given placebo who completed the study, all of whom had objective evidence of polyneuropathy, 7 out of 12 alrestatin-treated patients with subjective complaints at the time of entry reported improvement in symptoms, compared with 4 out of 13 placebo-treated patients reporting improvement after treatment. The mean scores for sensory impairment and discriminative sensation significantly improved in patients receiving the drug but not in those receiving placebo. Sensory thresholds, especially vibratory, improved on drug but not on placebo. Motor conduction velocities, especially in the ulnar nerve, also significantly improved with drug therapy. Galvanic skin response, a measure of autonomic function, improved with alrestatin, although not significantly. Using a summary of all the variables assessed in the evaluation of polyneuropathy in this study, the investigators concluded that signs of improvement occurred in 13 of the 14 patients treated with alrestatin, and suggested that treatment with an aldose reductase inhibitor might in the future be instituted early in the course of diabetes, "perhaps when the first clinical or even neurophysiological sign of polyneuropathy appears." Given the unknown effects of long-term inhibition of aldose reductase and the present uncertainty of the clinical benefits, such a prediction seems premature, if not ill advised.

A fourth study with alrestatin used the drug in a single-blind, nonrandomized, placebo-controlled crossover trial over a 4-month period in nine patients with severe, painful diabetic neuropathy.[4] Eight patients reported subjective improvement after 3 weeks of the drug, and the five who completed 8 weeks of treatment continued to feel improved. Three of these five reported return of pretrial level of symptoms after the subsequent 8 weeks of placebo. Four patients could not complete the study because of drug toxicity (rash, nausea, change in hepatic or renal function). Treatment with alrestatin had no objective effect on motor or sensory nerve function, as assessed by conduction velocities, latency, and amplitude.

The other published clinical trials have used Sorbinil, a spirohydantoin derivative developed by Pfizer, for the treatment of symptomatic diabetic neuropathy.[5-9] Perhaps the most widely known and cited of these studies, and probably the most encouraging, was that conducted in two medical centers with 39 patients who received Sorbinil for 9 weeks in a randomized, double-blind crossover trial.[5] Nerve conduction velocities in the peroneal motor, median motor, and median sensory nerves were greater during treatment with

Sorbinil, 250 mg/day, than during the placebo period, and conduction velocities for all three nerves declined significantly within 3 weeks after the drug was discontinued. Although the magnitude of the Sorbinil effect on conduction velocities was small, the investigators pointed out that there may be both reversible and irreversible components to diabetic neuropathy, that only the former may be amenable to treatment with aldose reductase inhibitors, and that irreversibility may be a function of time.

A subsequently published double-blind, placebo-controlled crossover trial examined the effect of Sorbinil, 200 mg daily for 4 weeks, in 13 patients ranging in age from 42 to 72 years, with a mean duration of diabetes of 17.3 years.[6] Response was assessed by tests of motor, sensory, and autonomic nerve function and by patient evaluation of pain severity and sleep duration. None of the patients noted subjective improvement with Sorbinil treatment, and the drug had no significant effect on motor or sensory nerve conduction velocities, vibration perception thresholds, or beat-to-beat variation in the heart rate as a measure of autonomic function. One patient had a toxic reaction, which manifested as fever, rash, and oral ulceration, that necessitated withdrawal from the study and that resolved after discontinuing the drug. The investigators noted that their patients were older, had a longer duration of diabetes, and had more severe clinical symptoms than did the patients in whom improvement in motor conduction velocity was observed in the larger clinical trial conducted in the two centers.

A third study with Sorbinil, performed in a double-blind, randomized, placebo-controlled manner, gave the drug (200 mg daily in two doses) to 15 patients ranging in age from 35 to 68 years who had chronic painful diabetic neuropathy.[7] Treatment was evaluated by subjective assessment of change in pain severity; clinical examination of tendon reflexes and of response to touch, vibration, pain, and temperature; motor and sensory nerve electrophysiology; and cardiovascular reflex tests of autonomic nerve function. Of the 12 patients who received Sorbinil for 27 consecutive days, 10 reported symptomatic improvement while taking the drug; subjective improvement persisted in four of five patients who received placebo during the subsequent 4 weeks. Sensory impairment seemed to worsen on the drug, but tendon reflexes seemed to improve. Sensory potentials in the sural nerve were greater during drug treatment, but there was no improvement in motor nerve conduction, and sensory nerve conduction may have deteriorated. The investigators speculated that subjective response in the absence of objective electrophysiologic evidence of improvement may indicate that the pain of diabetic neuropathy reflects metabolic rather than structural neural damage. Four patients in this study experienced adverse side effects while taking Sorbinil, consisting of an erythematous maculopapular rash associated with oropharyngeal involvement in three patients, transient leukopenia in two patients, fever in one patient, and cervical lymphadenopathy in another. The rashes completely resolved within 10 days after the drug was discontinued. However, adverse reactions to Sorbinil appear to be a significant problem. In a group of 45 patients exposed to the drug, 16 patients were noted to have drug-related problems.[17] In 11 of these patients, such problems took the form of a febrile illness variably associated with myalgia, lymphadenopathy, a maculopapular rash, transient neutropenia, thrombocytopenia, mild derangement of liver function tests, and a worsening of glycemic control. Although symptomatic recovery occurred within 10 days after discontinuing the drug, abnormalities in tests of liver function persisted in some patients.

In another study, 11 patients with severely painful diabetic neuropathy were treated with a single daily dose of Sorbinil, 250 mg, for periods ranging from 10 days to 5 weeks.[8] Eight of the patients received placebo, in single-blind fashion, for 4 days to 3 weeks, before taking the drug. Response was assessed by subjective rating on a pain scale, measurement of motor and sensory nerve conduction velocities, and evaluation of autonomic function by electrocardiographic monitoring of expiration/inspiration ratios of the heart rate. Eight patients reported moderate to marked relief of symptoms, and four patients with diabetic amyotrophy experienced improvement in proximal muscle strength. Motor and sensory conduction velocities improved in these four patients. Autonomic nerve function improved in six of seven patients in whom it was tested. However, a letter to the editor regarding this study noted that the patients who improved most had forms of painful neuropathy known to undergo spontaneous remission, and that of the total of 21 estimates of nerve conduction in individual patients, 38% showed deterioration during treatment.[18] No toxic effects of the drug were noted in this group of 11 patients, although a twelfth patient had a macular-erythematous rash on the sixth day of treatment and was withdrawn from the study.

The most recent publication describing results of treatment with Sorbinil reports on symptomatic responses and electrophysiologic and neurophysiologic parameters in 37 patients with diabetic neuropathy, 18 of whom received 50 mg daily and 19 of whom received 200 mg daily for 4 weeks.[9] Although the drug produced no significant effect in vibratory perception thresholds or on nerve conduction velocities or amplitudes, nine of the patients receiving 200 mg/day reported subjective improvement. Five of the patients treated with 50 mg/day, believed to be an inadequate therapeutic dose, also reported an improved sense of well-being while taking the drug. Retinal function, evaluated by adaptation to darkness with nyctometry, did not change during treatment with Sorbinil. No change in blood chemistries was observed, and there were no reported clinical side effects of the drug. The investigators noted that their patients were older (mean age 54 years) and had a longer mean duration of diabetes (17 years) than did the group of patients in the two-center study who showed improvement in nerve conduction velocity after receiving Sorbinil. The mean age of that group was 48 years, and the mean duration of diabetes was 9.8 years.

It appears that Sorbinil produces inconsistent, if any, improvement in objective parameters of nerve dysfunction, but can result in symptomatic improvement, the reasons for which are not entirely clear. Alleviation of symptoms and modest improvement in tests of nerve function, if they occur, are most likely to be seen in younger diabetic patients in whom the duration of diabetes and of neuropathic symptoms has not been unduly long. The question of toxicity, particularly desquamative dermatitis, is of concern to Pfizer, the Food and Drug Administration, and the physicians and patients who have used this agent, while the question of efficacy has cast doubt on the ultimate therapeutic role of aldose reductase inhibitors in general. Nevertheless, it is clear from the number of pharmaceutical houses developing aldose reductase inhibitors, undertaking programs for their clinical testing, and competitively racing for their market launch that discussion of these drugs will continue in the medical and lay news for some time to come.

References

1. Gabbay KH, Spack N, Loo S, et al: Aldose reductase inhibitors: Studies with alrestatin. *Metabolism* 1979; 28(suppl 1):471–476.

2. Culebras A, Alio J, Herrara JL, et al: Effect of an aldose reductase inhibitor on diabetic peripheral neuropathy. *Arch Neurol* 1981; 38:133–134.

3. Fagius J, Jameson S: Effects of aldose reductase inhibitor treatment in diabetic polyneuropathy—a clinical and neurophysiological study. *J Neurol Neurosurg Psychiatry* 1981; 44:999–1001.

4. Handlesman DJ, Turtle JR: Clinical trial of an aldose reductase inhibitor in diabetic neuropathy. *Diabetes* 1981; 30:459–464.

5. Judzewitsch RG, Jaspan JB, Polonsky KS, et al: Aldose reductase inhibitor improves nerve conduction velocity in diabetic patients. *N Engl J Med* 1983; 308:119–125.

6. Lewin IG, O'Brien IAD, Morgan MH, et al: Clinical and neurophysiological studies with the aldose reductase inhibitor, Sorbinil, in symptomatic diabetic neuropathy. *Diabetologia* 1984; 26:445–448.

7. Young RJ, Ewing DJ, Clarke BF: A controlled trial of Sorbinil, an aldose reductase inhibitor, in chronic painful diabetic neuropathy. *Diabetes* 1983; 32:938–942.

8. Jaspan J, Herold K, Maselli R, et al: Treatment of severe painful diabetic neuropathy with an aldose reductase inhibitor: Relief of pain and improved somatic and autonomic nerve function. *Lancet* 1983; 2:758–762.

9. Christensen JEJ, Varnek L, Gregersen G: The effect of an aldose reductase inhibitor (Sorbinil) on diabetic neuropathy and neural function of the retina. *Acta Neurol Scand* 1985; 71:164–167.

10. Fagius J, Jameson S: Treatment of diabetic polyneuropathy with an aldose reductase inhibitor—a clinical and neurophysiological study. *Acta Neurol Scand* 1980; 62(suppl 6):125A.

11. Gonzalez ER: Can aldose reductase inhibition ameliorate diabetic neuropathy? *JAMA* 1981; 248:1169–1170.

12. Hotta N, Kakuta H, Kimura M, et al: Experimental and clinical trial of aldose reductase inhibitor in diabetic neuropathy. *Diabetes* 1983; 32(suppl 1):98A.

13. Lehtinen JM, Hjvonen SK, Uusitupa M, et al: The effect of an aldose reductase inhibitor (Sorbinil) on diabetic neuropathy. *Diabetologia* 1984; 27:303A.

14. Proceedings of Aldose Reductase Inhibitor Symposium, London, 1984. *Diabetic Medicine*, vol 2, 1985.

15. Proceedings of Puerto Rico Conference on Sorbinil and Diabetic Complications. *Metabolism* (suppl), April 1986.

16. Koglin L, Clark C, Ryder S, et al: The results of the long-term open-label administration of ALREASE™ in the treatment of diabetic neuropathy. *Diabetes* 1985; 34(suppl 1):202A.

17. Martyn CN, Matthews DM, Popp-Snijders C, et al: Effects of Sorbinil treatment on erythrocytes and platelets of persons with diabetes. *Diabetes Care* 1985; 9:36–39.

18. Young RJ, Matthews DM, Clarke BF, et al: Aldose reductase inhibition for diabetic neuropathy (letter to the editor). *Lancet* 1983; 2:969.

Bibliography

Addison DJ, Garner A, Ashton N: Degeneration of intramural pericytes in diabetic retinopathy. *Br Med J* 1970; 1:264–266.

Akagi Y, Kador PF, Kuwabara T, et al: Aldose reductase localization in human retinal mural cells. *Invest Ophthalmol Vis Sci* 1983; 24:1516–1519.

Akagi Y, Yajima Y, Kador PF, et al: Localization of aldose reductase in the human eye. *Diabetes* 1984; 33:562–566.

Asbury AK, Aldridge H, Hershberg R, et al: Oculomotor palsy in diabetes mellitus: A clinico-pathologic study. *Brain* 1970; 93:555–566.

Ashton N: The blood-retinal barrier and vaso-glial relationships in retinal disease. *Trans Ophthalmol Soc UK* 1965; 85:199–230.

Attwood MA, Doughty CC: Purification and properties of calf liver aldose reductase. *Biochim Biophys Acta* 1974; 370:358–368.

Baba Y, Kai M, Kamada T, et al: Higher levels of erythrocyte membrane microviscosity in diabetes. *Diabetes* 1979; 28:1138–1140.

Barbosa J: Plasma *myo*-inositol in diabetics including patients with renal allografts. *Acta Diabetol Lat* 1978; 15:95–101.

Barnett PA, Gonzalez RG, Chylack LT, et al: The effect of oxidation on sorbitol pathway kinetics. *Diabetes* 1986; 35:426–432.

Barnes AJ, Locke P, Scudder PR, et al: Is hyperviscosity a treatable component of diabetic microcirculatory disease? *Lancet* 1971; 2:789–791.

Basak Halder A, Wolff S, Ting H-H, et al: An aldose reductase from the human erythrocyte. *Biochem Soc Trans* 1980; 8:644–645.

Behse F, Buchthal F, Carlson FI: Nerve biopsy and conduction studies in diabetic neuropathy. *J Neurol Neurosurg Psychiatry* 1977; 40:1072–1082.

Bell ME, Peterson RG, Eichberg J: Metabolism of phospholipids in peripheral nerve from rats with chronic streptozotocin-induced diabetes: Increased turnover of phosphatidyl-4,5-bisphosphate. *J Neurochem* 1982; 39:192–200.

Berti-Mattera L, Peterson R, Bell M, et al: Effect of hyperglycemia and its prevention by insulin treatment on the incorporation of [^{32}P] into polyphosphoinositides and other phospholipids in peripheral nerve of the streptozotocin diabetic rat. *J Neurochem* 1985; 45:1692–1698.

Beyer-Mears A, Cruz E: Reversal of diabetic cataract by Sorbinil, an aldose reductase inhibitor. *Diabetes* 1985; 34:15–21.

Beyer-Mears A, Cruz E, Dillon P, et al: Diabetic renal hypertrophy diminished by aldose reductase inhibition. *Fed Proc* 1983; 42:505.

Beyer-Mears A, Cruz E, Nicolas-Alexandre J, et al: Xanthone-2-carboxylic acid effect on lens growth, hydration and proteins during diabetic cataract development. *Arch Int Pharmacodyn* 1982; 259:166–176.

Beyer-Mears A, Cruz E, Nicholas-Alexandre J, et al: Sorbinil protection of lens protein components and cell hydration during diabetic cataract formation. *Pharmacology* 1982; 24:193–200.

Beyer-Mears A, Farnsworth PN: Diminished diabetic cataractogenesis by quercitin. *Exp Eye Res* 1979; 28:709–716.

Beyer-Mears A, Ku L, Cohen MP: Glomerular polyol accumulation in diabetes and its prevention by oral Sorbinil. *Diabetes* 1984; 33:604–607.

Beyer-Mears A, Nicolas-Alexandre J, Cruz E: Sorbinil inhibition of renal aldose reductase. *Fed Proc* 1983; 42:858.

Beyer-Mears A, Varagiannis E, Cruz E: Effect of Sorbinil on reversal of proteinuria. *Diabetes* 1985; 35(suppl 1):101A.

Bhuyan KC, Bhuyan DK, Katzin DM: Amizol-induced cataract and inhibition of lens catalase in rabbit. *Ophthalmic Res* 1973; 5:236–247.

Bischoff A: Morphology of diabetic neuropathy. *Horm Metab Res* 1980; 9(suppl):18–28.

Blankenship GW, Machemer R: Pars plana vitrectomy for the management of severe diabetic retinopathy: An analysis of results five years following surgery. *Ophthalmology (Rochester)* 1978; 85:553–559.

Boghosian RA, McGuiness ET: Affinity purification and properties of porcine brain aldose reductase. *Biochim Biophys Acta* 1979; 567:278–286.

Boghosian RA, McGuiness ET: Pig brain aldose reductase: A kinetic study using centrifugal fast analyzer. *Int J Biochem* 1981; 13:909–914.

Boot-Hanford R, Heath H: The effects of aldose reductase inhibitors on the metabolism of cultured monkey kidney epithelial cells. *Biochem Pharmacol* 1981; 30:3065–3069.

Bous F, Hockwin O, Ohrloff C, et al: Investigation on phosphofructokinase (EC 2.7.1.11) in bovine lens in dependence on age, topographic distribution and water soluble protein fractions. *Exp Eye Res* 1977; 24:383–389.

Branlant G: Properties of an aldose reductase from pig lens. *Eur J Biochem* 1982; 129:99–104.

Brightbill FS, Myers FL, Bresnick GN: Postvitrectomy keratopathy. *Am J Ophthalmol* 1978; 85:651–655.

Brismar T, Sima AAF: Changes in nodal function in nerve fibres of the spontaneously diabetic BB-Wistar rat: Potential clamp analysis. *Acta Physiol Scand* 1981; 113:499–506.

Broekhuyse RM: Changes in *myo*-inositol permeability in the lens due to cataractous conditions. *Biochem Biophys Acta* 1968; 163:269–272.

Brown MJ, Sumner AJ, Greene DA, et al: Distal neuropathy in experimental diabetes. *Ann Neurol* 1980; 8:168–178.

Brownlee M, Spiro RG: Glomerular basement membrane metabolism in the diabetic rat. *In vivo* studies. *Diabetes* 1979; 28:121–125.

Buzney SM, Frank RN, Varma SD, et al: Aldose reductase in retinal mural cells. *Invest Ophthalmol Vis Sci* 1977; 16:392–396.

Carandente O, Colombo R, Girardi AM, et al: Role of red cell sorbitol as determinant of reduced erythrocyte filtrability in insulin dependent diabetics. *Acta Diabetol Lat* 1982; 19(4):359–369.

Cataliotti R: Spectroscopic evidence of structural modifications in erythrocyte membranes of diabetic patients. *Stud Biophys* 1978; 73:199.

Chand D, El-Aguizy K, Richards RD, et al: Sugar cataracts in vitro: Implications of oxidative stress and aldose reductase I. *Exp Eye Res* 1982; 35:491–497.

Chang K, Rowold E, Marvel J, et al: Increased [125I]-albumin permeation of vessels in rats fed galactose. *Diabetes* 1985; 34(suppl 1):40A.

Chang K, Uitto J, Rowold EA, et al: Increased collagen cross-linkages in experimental diabetes. *Diabetes* 1980; 29:778–781.

Chaudhry PS, Cabrera J, Juliani HR, et al: Inhibition of human lens aldose reductase by flavinoids, sulindac and indomethacin. *Biochem Pharmacol* 1983; 32:1995–1998.

Chen HM, Chylack LT Jr: Factors affecting the rate of lactate production in rat lens. *Ophthalmic Res* 1977; 9:381–387.

Chiou SH, Chylack LT, Bunn HF, et al: Role of nonenzymatic glycosylation in experimental cataract formation. *Biochem Biophys Res Commun* 1980; 95:894–901.

Chochinov RH, Ullyot LE, Moorhouse JA: Sensory perception thresholds in patients with juvenile diabetes and their close relatives. *N Engl J Med* 1972; 286:1233–1237.

Christensen JEJ, Varnek L, Gregersen G: The effect of an aldose reductase inhibitor (Sorbinil) on diabetic neuropathy and neural function of the retina. *Acta Neurol Scand* 1985; 71:164–167.

Chylack LT, Cheng H-M: Sugar metabolism in the crystalline lens. *Surv Ophthalmol* 1978; 23:26–34.

Chylack LT Jr, Henriques HF, Cheng H-M, et al: Efficacy of alrestatin, an aldose reductase inhibitor, in human diabetic and nondiabetic lenses. *Ophthalmology* 1978; 86:1579.

Chylack LT Jr, Henriques H, Tung W: Inhibition of sorbitol production in human lenses by an aldose reductase inhibitor. *Invest Ophthalmol vis Sci* 1978; 17(ARVO Suppl):300.

Chylack LT, Kinoshita JH: Biochemical evaluation of a cataract induced in high glucose medium. *Invest Ophthalmol Vis Sci* 1969; 8:401–412.

Clark DL, Harnel FG, Queener SF: Changes in renal phospholipid fatty acids in diabetes mellitus; correlation with changes in adenylate cyclase activity. *Lipids* 1983; 18:696–705.

Clements RS: Diabetic neuropathy—Newer concepts in etiology. *Diabetes* 1979; 28:604–611.

Clements RS, Bell DHS: Diagnostic, pathogenetic and therapeutic aspects of diabetic neuropathy. *Spec Topics Endocrinol Metab* 1982; 3:1–43.

Clements RS, Diethelm AG: The metabolism of *myo*-inositol by the human kidney. *J Lab Clin Med* 1979; 93:210–219.

Clements RS, Morrison AD, Winegrad AI: Polyol pathway in aorta. Regulation by hormones. *Science* 1969; 166:1007–1008.

Clements RS, Reynertson RH, Starnes WS: *Myo*-inositol metabolism in diabetes mellitus. *Diabetes* 1974; 23:348A.

Clements RS, Reynertson R: *Myo*-inositol metabolism in diabetes mellitus. Effect of insulin treatment. *Diabetes* 1977; 26:215–221.

Clements RS, Stockard CR: Abnormal sciatic nerve *myo*-inositol metabolism in the streptozotocin-diabetic rat. Effects of insulin treatment. *Diabetes* 1980; 29:227–235.

Clements R, Winegrad AI: Purification of alditol NADP oxidoreductase from human placenta. *Biochem Biophys Res Commun* 1972; 47:1473–1480.

Cogan DG, Kinoshita JH, Kador PF, et al: Aldose reductase and complications of diabetes. *Ann Intern Med* 1984; 101:82–91.

Cogan DG, Toussaint D, Kuwabara T: Retinal vascular patterns. IV. Diabetic retinopathy. *Arch Ophthalmol* 1976; 66:366–372.

Cohen AH, Mampaso F, Zamboui L: Glomerular podocyte degeneration in human renal disease. *Lab Invest* 1977; 37:40–42.

Cohen MP: Glomerular basement membrane synthesis in streptozotocin diabetes, in Podolsky S, Viswanathan M (eds): *Secondary Diabetes, the Spectrum of the Diabetic Syndromes*, pp 541–551. New York, Raven Press, 1980.

Cohen MP: *Diabetes and Protein Glycosylation: Measurement and Biologic Relevance*. New York, Springer-Verlag, 1986.

Cohen MP, Dasmahapatra A, Shapiro E: Reduced glomerular sodium/potassium adenosine triphosphatase activity in acute streptozotocin diabetes. *Diabetes* 1985; 34:1071–1074.

Cohen MP, Dasmahapatra A, Wu VY: Deposition of basement membrane in vitro by normal and diabetic renal glomeruli. *Nephron* 1981; 27:146–151.

Cohen MP, Khalifa A: Renal glomerular collagen synthesis in streptozotocin diabetes. Reversal of increased basement membrane synthesis with insulin therapy. *Biochim Biophys Acta* 1977; 500:395–404.

Cohen MP, Klepser H: Binding of an aldose reductase inhibitor to renal glomeruli. *Biochem Biophys Res Commun* 1985; 129:530–535.

Cohen MP, Klepser H, Cua E: Effect of diabetes and Sorbinil treatment on phospholipid metabolism in rat glomeruli. *Biochim Biophys Acta* 1986; 876:226–232.

Cohen MP, Surma ML: [^{35}S]-Sulfate incorporation into glomerular basement membrane glycosaminoglycans is decreased in experimental diabetes. *J Lab Clin Med* 1981; 98:715–722.

Cohen MP, Surma ML: Effect of diabetes on in vivo metabolism of [^{35}S]-labeled glomerular basement membrane. *Diabetes* 1984; 33:8–12.

Cohen MP, Surma ML, Wu VY: In vivo biosynthesis and turnover of glomerular basement membrane in diabetic rats. *Am J Physiol* 1982; 242:F385–F389.

Cohen MP, Urdanivia E, Surma M, et al: Increased glycosylation of glomerular basement membrane collagen in diabetes. *Biochem Biophys Res Commun* 1980; 95:765–769.

Cohen MP, Wu VY: Identification of specific amino acids in diabetic glomerular basement membrane subject to nonenzymatic glycosylation in vivo. *Biochem Biophys Res Commun* 1984; 100:1549–1554.

Conrad SM, Doughty CC: Comparative studies on aldose reductase from bovine rat and human lens. *Biochim Biophys Acta* 1982; 708:348–357.

Cooper RA: Abnormalities of cell membrane fluidity in the pathogenesis of disease. *N Engl J Med* 1977; 297:371–377.

Corder CN, Braughler JM, Culp PA: Quantitative histochemistry of the sorbitol pathway in glomeruli and small arteries of human diabetic kidney. *Folia Histochem Cytochem (Krakow)* 1979; 17:137–146.

Corder CN, Collins JG, Brannon TS, et al: Aldose reductase and sorbitol dehydrogenase distribution in rat kidney. *J Histochem Cytochem* 1977; 25:1–8.

Cotlier E: *Myo*-inositol active transport by the crystalline lens. *Invest Opthalmol* 1970; 9:681–691.

Cotlier E: Aspirin effect on cataract formation in patients with rheumatoid arthritis alone or combined with diabetes. *Int Ophthalmol* 1981; 3:173–179.

Cotlier E, Fagadau W, Cicchetti DV: Methods for evaluation of medical therapy of senile and diabetic cataracts. *Trans Ophthalmol Soc UK* 1982; 102:416–422.

Cotlier E, Sharma YR, Niven T, et al: Distribution of salicylate in lens and intraocular fluids and its effect on cataract formation. *Am J Med* 1983; 75(6A):83–90.

Crabbe MJC, Basak Halder A: Affinity chromatography of bovine lens aldose reductase, and a comparison of some kinetic properties of the enzyme from lens and human erythrocyte. *Biochem Soc Trans* 1980; 8:194–195.

Crabbe MJC, Brown AJ, Peckar CO, et al: NADPH-oxidizing activity in lens and erythrocytes in diabetic and nondiabetic patients with cataract. *Br J Ophthalmol* 1983; 67:696–699.

Crabbe MJC, Freeman G, Halder AB, et al: The inhibition of bovine lens aldose reductase by Clinoril, its absorption into the human red cell and its effect on human red cell aldose reductase activity. *Ophthalmic Res* 1985; 17:85–89.

Crabbe MJC, Halder AB: Kinetic behavior under defined assay conditions for bovine lens aldose reductase. *Clin Biochem* 1979; 12:281–283.

Crabbe MJC, Halder AB, Peckar CO, et al: Erythrocyte glyceraldehyde-reductase levels in diabetes with retinopathy and cataracts. *Lancet* 1980; 2:1268–1270.

Creighton MO, Trevithick JR: Cortical cataract formation prevented by vitamin E and glutathione. *Exp Eye Res* 1979; 29:689–693.

Cromlish JA, Flynn TG: Pig muscle aldehyde reductase. Identity of pig muscle aldehyde reductase with pig lens aldose reductase and with low Km aldehyde reductase of pig brain and pig kidney. *J Biol Chem* 1983; 258:3583–3586.

Cromlish JA, Flynn TG: Purification and characterization of two aldose reductase isoenzymes from rabbit muscle. *J Biol Chem* 1983; 256:3416–3424.

Culebras A, Alio J, Herrera JL, et al: Effect of an aldose reductase inhibitor on diabetic peripheral neuropathy. *Arch Neurol* 1981; 38:133–134.

Cunha-Vaz J, DeAbreau JRF, Campos AJ, et al: Early breakdown of the blood-retinal barrier in diabetes. *Br J Ophthalmol* 1975; 59:649–656.

Cunha-Vaz JG, Mota C, Leite E, et al: Effect of aldose reductase inhibitors on the blood retinal barrier in early diabetic retinopathy. *Diabetes* 1985; 34(suppl 1): 109A.

Cunha-Vaz JG, Mota CC, Leite EC, et al: Effect of sulindac on the permeability of the blood retinal barrier in early diabetic retinopathy. *Arch Ophthalmol* 1985; 103:1307–1311.

Cunha-Vaz JG, Mota CC, Leite EC, et al: Effect of Sorbinil on blood-retinal barrier in early diabetic retinopathy. *Diabetes* 1986; 35:574–578.

Das B, Hair GA, Srivastava SK: Activated and unactivated forms of aldose reductase and its role in diabetic complications. *Fed Proc* 1985; 44:1391(A).

Das B, Srivastava SK: Purification and properties of aldose reductase and aldehyde reductase II from human erythrocyte. *Arch Biochem Biophys* 1985; 238:670–679.

Das B, Srivastava SK: Activation of aldose reductase from tissues. *Diabetes* 1985; 34:1145–1151.

Datiles M, Fukui H, Kuwabara T, et al: Galactose cataract prevention with Sorbinil, an aldose reductase inhibitor: A light microscopic study. *Invest Ophthalmol Vis Sci* 1982; 22:174–179.

Datiles MD, Kador PF, Fukui HN, et al: Corneal re-epithelialization in galacto-semic rats. *Invest Ophthalmol Vis Sci* 1983; 24:563–569.

Daughaday WH, Larner JL: The renal excretion of inositol in normal and diabetic human beings. *J Clin Invest* 1954; 33:326–332.

Davies PD, Duncan G, Pynsent PB, et al: Aqueous humor glucose concentration in cataract patients and its effect on the lens. *Exp Eye Res* 1984; 39:605–609.

Dische Z, Zil H: Studies on the oxidation of cysteine to cystine in lens protein dur-ing cataract formation. *Am J Ophthalmol* 1951; 34:104–113.

Dons RF, Doughty CC: Isolation and characterization of aldose reductase from calf brain. *Biochim Biophys Acta* 1976; 452:1–12.

Dormandy JA, Hoare E, Colley U, et al: Clinical, haemodynamic, rheological and biochemical findings in 126 patients with intermittent claudication. *Br Med J* 1973; IV:576–581.

Dormandy JA, Hoare E, Dhatab AH, et al: Prognostic significance of rheological and biochemical findings in patients with intermittent claudication. *Br Med J* 1973; IV:581–583.

Doughty CC, Lee S-M, Conrad S, et al: Kinetic mechanism and structural proper-ties of lens aldose reductase, in Weiner H, Wermuth B (eds): *Enzymology of Car-bonyl Metabolism: Aldehyde Dehydrogenase and Aldo/Keto Reductase*, pp 223–242. New York, Alan R. Liss, 1982.

Dreyfus PM, Hakim S, Adams RD: Diabetic ophthalmoplegia: Report of a case with postmortem study and comments on vascular supply of human oculomotor nerve. *Arch Neurol Psychiatry* 1957; 77:337–349.

Duke-Elder WS: Changes in refraction in diabetes mellitus. *Br J Ophthalmol* 1925; 9:167–187.

Dvornik D, Simard-Duquesne N, Kraml M, et al: Polyol accumulation in galacto-semic and diabetic rats: Control by an aldose reductase inhibitor. *Science* 1973; 182:1146–1148.

Dyck PJ, Lamberg EH, Windebank AJ, et al: Acute hyperosmolar hyperglycemia causes axonal shrinkage and reduced nerve conduction velocity. *Exp Neurol* 1981; 71:507–514.

Dyck PJ, Sherman WR, Hallcher LM, et al: Human diabetic endoneurial sorbitol, fructose and *myo*-inositol related to sural nerve morphometry. *Ann Neurol* 1980; 8:590–596.

Eck MG, Wynn JO, Carter WJ, et al: Fatty acid desaturation in experimental dia-betes mellitus. *Diabetes* 1979; 28:479–485.

El-Aguizy HK, Richards RD, Varma SD: Sugar cataracts in mongolian gerbil (*Meri-ones unguiculatus*). *Exp Eye Res* 1983; 36:839–844.

Engerman RL, Kern TS: Experimental galactosemia produces diabetic-like retinopathy. *Diabetes* 1984; 33:97–100.

Fagius J: Microneurographic findings in diabetic polyneuropathy with special reference to sympathetic nerve activity. *Diabetologia* 1982; 23:415–420.

Fagius J, Jameson S: Treatment of diabetic polyneuropathy with an aldose reductase inhibitor—a clinical neurophysiological study. *Acta Neurol Scand* 1980; 62(suppl 6):125.

Fagius J, Jameson S: Effects of aldose reductase inhibitor treatment in diabetic polyneuropathy—a clinical and neurophysiological study. *J Neurol Neurosurg Psychiatry* 1981; 44:999–1001.

Farese RV: Phosphoinositide metabolism and hormone action. *Endocr Rev* 1983; 4:78–95.

Faulborn J, Conway BP, Machemer R: Surgical complications of pars plana vitreous surgery. *Ophthalmology (Rochester)* 1978; 85:116–125.

Feldman HB, Szczepanik PA, Harne P, et al: Stereospecificity of the hydrogen transfer catalyzed by human placental aldose reductase. *Biochim Biophys Acta* 1977; 480:14–20.

Finegold D, Lattimer SA, Nolle S, et al: Polyol pathway activity and *myo*-inositol metabolism. A suggested relationship in the pathogenesis of diabetic neuropathy. *Diabetes* 1983; 32:988–992.

Finegold DN, Nolle SS, Lattimer S, et al: Alterations in renal ouabain sensitive Na/K-ATPase activity in streptozotocin diabetes. *Clin Res* 1984; 32:395A.

Foulks GN, Thoft RA, Perry HD, et al: Factors related to corneal epithelial complications after closed vitrectomy in diabetics. *Arch Ophthalmol* 1979; 97:1076–1078.

Friend J, Snip RC, Kiorpes TC, et al: Insulin insensitivity and sorbitol production of the normal rabbit corneal epithelium in vitro. *Invest Ophthalmol Vis Sci* 1980; 19:913–919.

Fukuma M, Carpentier J-L, Orci L, et al: An alteration in internodal myelin membrane structure in large sciatic nerve fibres in rats with acute streptozotocin diabetes and impaired nerve conduction velocity. *Diabetologia* 1978; 15:65–72.

Fukushi H, Merola L, Kinoshita JH: Altering the course of cataracts in diabetic rats. *Invest Ophthalmol Vis Sci* 1980; 19:313–315.

Fukushi H, Merola LO, Tanaka M, et al: Reepithelialization of denuded corneas in diabetic rats. *Exp Eye Res* 1980; 31:611–621.

Gabbay KH: Purification and immunological identification of bovine retinal aldose reductase. *Isr J Med Sci* 1972; 8:1626–1628.

Gabbay KH: The sorbitol pathway and complications of diabetes. *N Engl J Med* 1973; 288:831–836.

Gabbay KH: Hyperglycemia, polyol metabolism and complications of diabetes. *Annu Rev Med* 1975; 26:521–536.

Gabbay KH, Cathcart ES: Purification and immunologic identification of aldose reductases. *Diabetes* 1974; 23:460–468.

Gabbay KH, Kinoshita JH: Growth hormone, sorbitol, and diabetic capillary disease. *Lancet* 1971; 1:913.

Gabbay KH, Merola LO, Field RA: Sorbitol pathway: Presence in nerve and cord with substrate accumulation in diabetes. *Science* 1966; 151:209–210.

Gabbay KH, O'Sullivan JB: The sorbitol pathway in diabetes and galactosemia: Enzyme and substrate localization. *Diabetes* 1968; 17:239–243.

Gabbay KH, O'Sullivan JB: The sorbitol pathway in diabetes and galactosemia: Enzyme and substrate localization and changes in the kidney. *Diabetes* 1968; 17:300A.

Gabbay KH, Snider JJ: Nerve conduction defect in galactose-fed rats. *Diabetes* 1972; 21:295–300.

Gabbay KH, Spack N, Loo S, et al: Aldose reductase inhibitors: studies with alrestatin. *Metabolism* 1979; 28(suppl 1):471–476.

Garner MH, Spector A: Selective oxidation of cysteine and methionine in normal cataractous lens. *Proc Natl Acad Sci USA* 1980; 77:1274–1277.

Gillon KRW, Hawthorne JN: Sorbitol, inositol and nerve conduction in diabetes. *Life Sci* 1983; 32:1943–1947.

Gillon KRW, Hawthorne JN, Tomlinson DR: *Myo*-inositol and sorbitol metabolism in relation to peripheral nerve function in experimental diabetes in the rat: The effect of aldose reductase inhibition. *Diabetologia* 1983; 25:365–371.

Goil MM, Harpur RP: Aldose reductase and sorbitol dehydrogenase in the muscle of *Ascaris suum* (Nematoda). *Parasitology* 1978; 77:97–102.

Goldfarb SA, Simmons DA, Kern E: Amelioration of glomerular hyperfiltration in acute experimental diabetes by dietary *myo*-inositol and by an aldose reductase inhibitor. *Clin Res* 1986; 34:725A.

Gonzalez ER: Can aldose reductase inhibition ameliorate diabetic neuropathy? *JAMA* 1981; 246:1169–1170.

Gonzalez AM, Sochor M, McLean P: Effect of experimental diabetes on glycolytic intermediates and regulation of phosphofructokinase in rat lens. *Biochem Biophys Res Commun* 1980; 95:1173–1179.

Gonzalez AM, Sochor M, McLean P: The effect of an aldose reductase inhibitor (Sorbinil) on the level of metabolites in lenses of diabetic rats. *Diabetes* 1983; 32:482–485.

Gonzalez AM, Sochor M, Rowles PM, et al: Sequential biochemical and structural changes occurring in rat lens during cataract formation in experimental diabetes. *Diabetologia* 1981; 21:5.

Goosey JD, Zigler JS Jr, Kinoshita JH: Cross-linking of lens crystallins in a photodynamic system. A singlet oxygen mediated process. *Science* 1980; 208:1278–1279.

Grachetti A: Axoplasmol transport of noradenaline in the sciatic nerves of spontaneously diabetic mice. *Diabetologia* 1979; 16:191–199.

Graf RJ, Halter JB, Halar E, et al: Nerve conduction abnormalities in untreated maturity-onset diabetes: Relation to levels of fasting plasma glucose and glycosylated hemoglobin. *Ann Intern Med* 1979; 90:298–303.

Greene DA, DeJesus PV, Winegrad AI: Effects of insulin and dietary *myo*-inositol on impaired peripheral motor nerve conduction velocities in acute streptozotocin diabetes. *J Clin Invest* 1975; 55:1326–1336.

Greene D, Lattimer SA: Sodium and energy-dependent uptake of *myo*-inositol by rabbit peripheral nerve. *J Clin Invest* 1982; 70:1009–1018.

Greene DA, Lattimer SA: Impaired rat sciatic nerve sodium-potassium ATPase in acute streptozotocin diabetes and its correction by dietary *myo*-inositol supplementation. *J Clin Invest* 1983; 72:1058–1063.

Greene DA, Lattimer SA: Impaired energy utilization and sodium-potassium ATPase in diabetic peripheral nerve. *Am J Physiol* 1984; 246:E311–E318.

Greene DA, Lattimer SA: Action of sorbinil in diabetic peripheral nerve. Relationship of polyol (sorbitol) pathway inhibition to a *myo*-inositol mediated defect in sodium-potassium ATPase activity. *Diabetes* 1984; 33:712–716.

Greene DA, Lattimer SA: Protein kinase C agonists acutely normalize ouabain-inhibitable respiration in diabetic rabbit nerve. *Diabetes* 1986; 35:242–245.

Greene DA, Lattimer SA, Sima AAF: Acute paranodal nerve fiber swelling and conduction slowing in the insulin-deficient BB rat reflects *myo*-inositol depletion and Na/K-ATPase deficiency rather than sorbitol accumulation. *Clin Res* 1986; 34:683A.

Greene DA, Lewis RA, Lattimer SA, et al: Selective effects of *myo*-inositol administration on sciatic and tibial motor nerve conduction parameters in the streptozotocin-diabetic rat. *Diabetes* 1982; 31:573–578.

Greene DA, Winegrad AI: In vitro studies of the substrates for energy production and the effects of insulin on glucose utilization in the neural components of peripheral nerve. *Diabetes* 1979; 28:878–887.

Greene DA, Winegrad AI: Effects of acute experimental diabetes on composite energy metabolism in peripheral nerve axons and Schwann cells. *Diabetes* 1981; 30:967–974.

Halder AB, Crabbe MJCC: Inhibition of aldose reductase by phenylglyoxal, diethylpyrocarbonate and thiol modifiers. *Biochem Soc Trans* 1982; 10:401–403.

Halder AB, Crabbe MJC: Bovine lens aldehyde reductase (aldose reductase). Purification, kinetics and mechanism. *Biochem J* 1984; 219:33–39.

Halder AB, Wolff S, Ting H-H, et al: An aldose reductase from human erythrocyte. *Biochem Soc Trans* 1980; 8:644–645.

Hamlett YC, Heath H: The accumulation of fructose-1-phosphate in the diabetic rat retina. *IRCS Med Sci* 1977; 5:510.

Handlesman DJ, Turtle JR: Clinical trial of an aldose reductase inhibitor in diabetic neuropathy. *Diabetes* 1981; 30:459–464.

Hanker JS, Ambrose WW, Yates PE, et al: Peripheral neuropathy in mouse hereditary diabetes mellitus. I. Comparison of neurologic, histologic and morphometric parameters with dystonic mice. *Acta Neuropathol (Berl)* 1980; 51:145–153.

Hastein T, Velle W: Placental aldose reductase activity and foetal blood fructose during bovine pregnancy. *J Reprod Fertil* 1968; 15:47–52.

Hastein T, Velle W: Purification and properties of aldose reductase from the placenta and seminal vesicle of the sheep. *Biochim Biophys Acta* 1969; 178:1–10.

Hawthorne JN, Pickard MP, Griffin HD: Phosphatidylinositol, triphosphoinositide and synaptic transmission, in Wells WW, Eisenberg F (eds): *Cyclitols and Polyphosphoinositides*, pp 145–151. New York, Academic Press, 1978.

Hayman S, Kinoshita JH: Isolation and properties of lens aldose reductase. *J Biol Chem* 1965; 240:877–882.

Hayman S, Lou MF, Merola LO, et al: Aldose reductase activity in the lens and other tissues. *Biochim Biophys Acta* 1966; 128:474–482.

Heath H, Kang SS, Philippou D: Glucose, glucose-6-phosphate, lactate and pyruvate content of the retina, blood and liver of streptozotocin-diabetic rats fed sucrose- or starch-rich diets. *Diabetologia* 1975; 11:57–62.

Heath H, Hamlett YC: The sorbitol pathway: Effect of streptozotocin induced diabetes and the feeding of a sucrose-rich diet on glucose, sorbitol and fructose in the retina, blood and liver of rats. *Diabetologia* 1976; 12:43–46.

Henrickson HS, Reinertsen JL: Phosphoinositide interconversion: A model for control of Na^+ and K^+ permeability in the nerve axon. *Biochem Biophys Res Commun* 1971; 44:1258–1264.

Hermann RK, Kador PF, Kinoshita JH: Rat lens aldose reductase: Rapid purification and comparison with human placental aldose reductase. *Exp Eye Res* 1983; 37:467–474.

Hers HG: Le mécanisme de la formation du fructose séminal et du fructose foetal. *Biochim Biophys Acta* 1960; 37:127–138.

Hers HG: Le mécanisme de la transformation de glucose en fructose par les vésicules séminales. *Biochim Biophys Acta* 1956; 22:202–203.

Hoare EM, Barnes AJ, Dormandy JA: Abnormal blood viscosity in diabetes mellitus and retinopathy. *Biorheology* 1978; 13:21.

Hockwin O, Bergeder HD, Kaiser L: Über die Galaktosekatarakt junger Ratten nach Ganzkorperröntgenbestrahlung. *Ber Dtsch Ophthalmol Ges* 1967; 68:135–139.

Hockwin O, Bergeder HD, Ninnemann U, et al: Untersuchungen zur Latenzzeit der Galaktosekatarakt von Ratten. Einfluss von Röntgenbestrahlung und Diätbeginn bei verschieden alten Tieren. *Graefes Arch Klin Ophthalmol* 1974; 189:171–178.

Hoffman PL, Wermuth B, von Wartburg J-P: Human brain aldehyde reductases: Relationship to succinic semialdehyde reductase and aldose reductase. *J Neurochem* 1980; 35:354–366.

Hogan MJ, Feeney L: The ultrastructure of the retinal vessels. II. The small vessels. *J Ultrastruct Res* 1963; 9:29–46.

Hollows JC, Schofield PJ, Williams JF, et al: The effect of an unsaturated fat-diet on cataract formation in streptozotocin-induced diabetic rats. *Br J Nutr* 1976; 36:161–177.

Hothersall JS, McLean P: Effect of diabetes and insulin on phosphatidylinositol synthesis in rat sciatic nerve. *Biochem Biophys Res Commun* 1979; 88:477–484.

Hotta N, Kakuta H, Fukasawa H, et al: Aldose reductase inhibitor and fructose-rich diet: Its effect on the development of diabetic retinopathy. *Diabetes* 1984; 33(suppl 1):199A.

Hotta N, Kakuta H, Kimura M, et al: Experimental and clinical trial of aldose reductase inhibitor in diabetic neuropathy. *Diabetes* 1983; 32(suppl 1):98A.

Hu T-S, Datiles M, Kinoshita JH: Reversal of galactose cataract with Sorbinil in rats. *Invest Ophthalmol Vis Sci* 1983; 24:640–644.

Huehns ER: Disorders of carbohydrate metabolism in the red blood corpuscle. *Clin Endocrinol Metab* 1976; 5:651–674.

Hutton JC, Schofield PJ, Williams JF, et al: Sorbitol metabolism in the retina: Accumulation of pathway intermediates in streptozotocin induced diabetes in the rat. *Aust J Exp Biol Med Sci* 1974; 52:361–373.

Hutton JC, Schofield PJ, Williams JF, et al: The localization of sorbitol pathway activity in the rat renal cortex and its relationship to the pathogenesis of the renal complications of diabetes mellitus. *Aust J Exp Biol Med Sci* 1975; 53:49–57.

Hutton JC, Schofield PJ, Williams JF, et al: The failure of aldose reductase inhibitor 3,3'-tetramethylene glutaric acid to inhibit in vivo sorbitol accumulation in lens and retina in diabetes. *Biochem Pharmacol* 1974; 23:2991–2998.

Hutton JC, Williams JF, Schofield PJ, et al: Polyol metabolism in monkey-kidney epithelial-cell cultures. *Eur J Biochem* 1974; 49:347–353.

Hyndiuk RA, Kazarian EL, Schultz RO, et al: Neurotrophic corneal ulcers in diabetes mellitus. *Arch Ophthalmol* 1977; 95:2193–2196.

Inagaki K, Miwa I, Yashiro T, et al: Inhibition of aldose reductases from rat and bovine lenses by hydantoin derivatives. *Chem Pharmacol Bull* 1984; 30:3244–3254.

Inagaki K, Miwa I, Okuda J: Affinity purification and glucose specificity of aldose reductase in bovine lens. *Arch Biochem Biophys* 1979; 216:337–344.

Ishikawa T: Fine structure of retinal vessels in man and the macaque monkey. *Invest Ophthalmol* 1963; 2:1–15.

Jacobson M, Sharma YR, Cotlier E, et al: Diabetic complications in lens and nerve and their prevention by sulindac or Sorbinil: Two novel aldose reductase inhibitors. *Invest Ophthalmol Vis Sci* 1983; 24:1426–1429.

Jaspan J, Herold K, Maselli R, et al: Treatment of severe painful diabetic neuropathy with an aldose reductase inhibitor: Relief of pain and improved somatic and autonomic nerve function. *Lancet* 1983; 2:758–762.

Jakobsen J: Axon dwindling in early experimental diabetes. I. A study of cross-sectioned nerves. *Diabetologia* 1976; 12:539–546.

Jakobsen J: Peripheral nerves in early experimental diabetes. Expansion of the endoneurial space as a cause of increased water content. *Diabetologia* 1978; 14:113–119.

Jakobsen J: Early and preventable changes of peripheral nerve structure and function in insulin-deficient diabetic rat. *J Neurol Neurosurg Psychiatry* 1979; 42:509–518.

Jakobsen J, Sidenius P: Decreased axonal transport of structural proteins in streptozotocin-diabetic rats. *J Clin Invest* 1980; 66:292–297.

Jedziniak JA, Chylack LT, Chen H-M, et al: The sorbitol pathway in the human lens: Aldose reductase and polyol dehydrogenase. *Invest Ophthalmol Vis Sci* 1981; 20:314–326.

Jedziniak JA, Kinoshita JH: Activators and inhibitors of lens aldose reductase. *Invest Ophthalmol* 1971; 10:357–366.

Jeffreys JGR, Palmano KP, Sharma AK, et al: Influence of dietary *myo*-inositol on nerve conduction and inositol phospholipids in normal and diabetic rats. *J Neurol Neurosurg Psychiatry* 1978; 41:333–339.

Johnson PC: Thickening of the human dorsal root ganglion perineurial cell basement membrane in diabetes mellitus. *Muscle Nerve* 1983; 6:561–565.

Johnson PC, Brendel K, Meezan E: Human diabetic perineurial cell basement membrane thickening. *Lab Invest* 1981; 44:165–170.

Jones DB: Correlative scanning and transmission electron microscopy of glomeruli. *Lab Invest* 1977; 37:569–578.

Jones DB: SEM of human and experimental renal disease. *Scan Electron Microsc* 1979; 3:679–689.

Judzewitsch RG, Jaspan JB, Polonsky KS, et al: Aldose reductase inhibitor improves nerve conduction velocity in diabetic patients. *N Engl J Med* 1983; 308:119–125.

Kador PF, Carper D, Kinoshita JH: Rapid purification of human placental aldose reductase. *Anal Biochem* 1981; 114:53–58.

Kador PF, Goosey JD, Sharpless NE, et al: Stereospecific inhibition of aldose reductase. *Eur J Med Chem* 1981; 16:293–298.

Kador PF, Kinoshita JH, Tung WH, et al: Differences in the susceptibility of aldose reductase to inhibition. *Invest Ophthalmol Vis Sci* 1980; 19:980–982.

Kador PF, Merola LO, Kinoshita JH: Differences in the susceptibility of aldose reductase to inhibition. *Doc Ophthalmol Proc Ser* 1979; 18:117–124.

Kador PF, Sharpless NE: Pharmacophor requirements of the aldose reductase inhibitor site. *Mol Pharmacol* 1983; 24:521–531.

Kador PF, Sharpless NE: Structure-activity studies of aldose reductase inhibitors containing the 4-oxo-4H-chromen ring system. *Biophys Chem* 1978; 8:81–85.

Kador PF, Sharpless NE, Goosey JD: Aldose reductase inhibition by anti-allergy compounds, in Weiner H, Wermuth B (eds): *Enzymology of Carbonyl Metabolism: Aldehyde Dehydrogenase and Aldo/Keto Reductase*, pp 243–259. New York, Alan R. Liss, 1982.

Kador PF, Zigler S, Kinoshita JH: Alterations of lens protein synthesis in galactosemic rats. *Invest Ophthalmol Vis Sci* 1979; 18:696–702.

Kadoya K, Hashi H, Yui MNH, et al: Influence of aldose reductase inhibitor on peroxidation reaction in the lens of streptozotocin diabetic rats. *Nippon Ganka Kiyo* 1983; 34:2172–2176.

Kahn HA, Liebowitz HM, Ganley JP, et al: The Framingham eye study. II. Association of ophthalmic pathology with single variables previously measured in the Framingham study. *Am J Epidemiol* 1977; 106:33–41.

Kamada T, Otsuji S: Low-levels of erythrocyte membrane fluidity in diabetic patients. A spin label study. *Diabetes* 1983; 32:585–591.

Kanski JJ: Anterior segment complications of retinal photocoagulation. *Am J Ophthalmol* 1975; 79:424–427.

Kanwar YS, Rosenzweig LJ, Linker A, et al: Decreased de novo synthesis of glomerular proteoglycans in diabetes: Biochemical and autoradiographic evidence. *Proc Natl Acad Sci USA* 1983; 80:2272–2275.

Kaplan SA, Lee W-NP, Scott ML: Glucose inhibits *myo*-inositol transport and phosphatidylinositol formation in adipocytes. *Diabetes* 1985; 34(suppl 1):183A.

Keller H-W, Stinnesbeck TH, Hockwin O, et al: Investigations on the influence of whole body X-irradiation on the activity of rat lens aldose reductase (E.C.1.1.1.21). *Graefes Arch Klin Ophthalmol* 1981; 215:181–186.

Kennedy A, Frank RN, Varma SD: Aldose reductase activity in retinal and cerebral microvessels and cultured vascular cells. *Invest Ophthalmol Vis Sci* 1983; 24:1250–1258.

Kern TS, Engerman RL: Distribution of aldose reductase in ocular tissues. *Exp Eye Res* 1981; 33:175–182.

Kern TS, Engerman RL: Immunohistochemical distribution of aldose reductase. *Histochem J* 1982; 14:507–515.

Kern TS, Engerman RL: Hexitol production by canine retinal microvessels. *Invest Ophthalmol Vis Sci* 1985; 26:382–384.

Khalifa A, Cohen MP: Glomerular protocollagen lysyl hydroxylase activity in streptozotocin diabetes. *Biochim Biophys Acta* 1975; 386:332–339.

Kikkawa R, Hatanaka I, Yasuda H, et al: Effect of a new aldose reductase inhibitor, (E)-3-carboxymethyl-5[(2E)-methyl-3-phenylpropenylidene] rhodanine (ONO-2235) on peripheral nerve disorders in streptozotocin-diabetic rats. *Diabetologia* 1984; 24:290–292.

Kilzer P, Chang K, Marvel J, et al: Albumin permeation of new vessels is increased in diabetic rats. *Diabetes* 1985; 34:333–336.

Kinoshita JH: Cataracts in galactosemia. *Invest Ophthalmol* 1965; 4:786–799.

Kinoshita JH: Mechanisms initiating cataract formation. *Invest Ophthalmol* 1974; 13:713–724.

Kinoshita JH, Barber GW, Merola LD, et al: Changes in the levels of free amino acids and *myo*-inositol in the galactose-exposed lens. *Invest Ophthalmol* 1969; 8:625–632.

Kinoshita JH, Dvornik D, Kraml M, et al: The effect of an aldose reductase inhibitor on the galactose-exposed rabbit lens. *Biochim Biophys Acta* 1968; 158:472–475.

Kinoshita JH, Fukushi S, Kador P, et al: Aldose reductase in diabetic complications of the eye. *Metabolism* 1979; 28(suppl 1):462–469.

Kinoshita JH, Futterman S, Satoh K, et al: Factors affecting the formation of sugar alcohols in ocular lens. *Biochim Biophys Acta* 1963; 74:340–350.

Kinoshita JH, Merola LO: Hydration of the lens during the development of galactose cataract. *Invest Ophthalmol* 1964; 3:577-584.

Kinoshita JH, Merola LO, Dikmak E: The accumulation of dulcitol and water in rabbit lens incubated with galactose. *Biochim Biophys Acta* 1962; 62:176-178.

Kinoshita JH, Merola LO, Dikmak E: Osmotic changes in experimental galactose cataracts. *Exp Eye Res* 1962; 1:405-410.

Kinoshita JH, Merola LO, Hayman S: Osmotic effects on the amino acid-concentrating mechanism in the rabbit lens. *J Biol Chem* 1965; 240:313-315.

Kinoshita JH, Merola LO, Satoh K, et al: Osmotic changes caused by the accumulation of dulcitol in the lenses of rats fed with galactose. *Nature (Lond)* 1962; 194:1085-1087.

Kinoshita JH, Merola O, Tung B: Changes in cation permeability in the galactose-exposed rabbit lens. *Exp Eye Res* 1968; 7:80-90.

Koglin L, Clark C, Ryder S, et al: The result of the long-term open-label administration of ALREDASE™ in the treatment of diabetic neuropathy. *Diabetes* 1985; 34(suppl 1):202A.

Krupin T, Waltman SR, Oestrich C, et al: Vitreous fluorophotometry in juvenile-onset diabetes mellitus. *Arch Ophthalmol* 1978; 96:812-814.

Krupin T, Waltman SR, Szewczyk P, et al: Fluorometric studies on the blood retinal barrier in experimental animals. *Arch Ophthalmol* 1982; 100:631-634.

Ku DD, Meezan E: Increased renal tubular sodium pump and Na,K-adenosine triphosphatase in streptozotocin-diabetic rats. *J Pharmacol Exp Ther* 1984; 229:664-670.

Kuwabara T, Cogan DG: Retinal vascular patterns. VI. Mural cells of the retinal capillaries. *Arch Ophthalmol* 1963; 69:492-502.

Kuwabara T, Kinoshita JH, Cogan DC: Electron microscopic study of galactose-induced cataract. *Invest Ophthalmol* 1969; 8:133-149.

Lee SM, Schade SZ, Doughty CC: Aldose reductase, NADPH and NADP$^+$ in normal, galactose-fed and diabetic rat lens. *Biochim Biophys Acta* 1985; 841:247-253.

LeFevre PG, Davis R: Active transport into the human erythrocyte: Evidence from comparative kinetics and composition among monosaccharides. *J Gen Physiol* 1951; 34:515-524.

Lehtinen JM, Hjönen SK, Uusitupa M, et al: The effect of an aldose reductase inhibitor (Sorbinil) on diabetic neuropathy. *Diabetologia* 1984; 27:303A.

Lerner BC, Varma SD, Richards RD: Polyol pathway metabolites in human cataracts. *Arch Ophthalmol* 1984; 102:917-920.

Lewin IG, O'Brien IAD, Morgan MH, et al: Clinical and neurophysiological studies with the aldose reductase inhibitor, Sorbinil, in symptomatic diabetic neuropathy. *Diabetologia* 1984; 26:445-448.

Li W, Chan LS, Khatami M, et al: Inhibition of *myo*-inositol uptake in cultured bovine retinal capillary pericytes by D-glucose: Reversal by Sorbinil. *Invest Ophthalmol Vis Sci* 1985; 26(suppl):335.

Li W, Chan LS, Khatami M, et al: Characterization of glucose transport by bovine retinal capillary pericytes in culture. *Exp Eye Res* 1985; 41:191-199.

Li W, Khatami M, Rockey JH: The effects of glucose and an aldose reductase inhibitor on the sorbitol content and collagen synthesis of bovine retinal capillary pericytes in culture. *Exp Eye Res* 1985; 40:439-444.

Li W, Shen S, Khatami M, et al: Stimulation of retinal capillary pericyte protein and collagen synthesis in culture by high-glucose concentration. *Diabetes* 1984; 33:785-789.

Lo C-S, August TR, Liberman UA, et al: Dependence of renal (Na/K-ATPase)-adenosine triphosphatase activity on thyroid status. *J Biol Chem* 1976; 251: 7826-7833.

Low PA, Dyck PJ, Schmelzer JD: Mammalian peripheral nerve sheath has unique responses to chronic elevations of endoneurial fluid pressure. *Exp Neurol* 1980; 70:300-306.

Low PA, Dyck PJ, Schmelzer JD: Chronic elevation of endoneurial fluid pressure is associated with low-grade fiber pathology. *Muscle Nerve* 1982; 5:162-165.

Ludvigson MA, Sorenson RL: Immunohistochemical localization of aldose reductase. I. Enzyme purification and antibody preparation—localization in peripheral nerve, artery and testis. *Diabetes* 1980; 29:438-449.

Ludvigson MA, Sorenson RL: Immunohistochemical localization of aldose reductase. II. Rat eye and kidney. *Diabetes* 1980; 29:450-459.

MacGregor LC, Matschinsky FM: Treatment with aldose reductase inhibitor or with *myo*-inositol arrests deterioration of the electroretinogram of diabetic rats. *J Clin Invest* 1985; 76:887-889.

MacGregor LC, Matschinsky FM: Correlation of biochemical and electrophysiological abnormalities in retinas of experimentally diabetic animals. *Diabetes* 1985; 34(suppl 1):13A.

MacGregor LC, Rosecan LR, Laties AM, et al: Microanalysis of total lipid, glucose, sorbitol, and *myo*-inositol in individual retinal layers of normal and alloxan diabetic rabbits. *Diabetes* 1984; 33:89A.

McMillan DE, Utterback NG, La Puma J: Reduced erythrocyte deformability in diabetes. *Diabetes* 1978; 27:895-901.

Malone J, Knox G, Benford S, et al: Red cell sorbitol—an indicator of diabetic control. *Diabetes* 1980; 29:861-864.

Malone JI, Knox G, Harvey C: Sorbitol accumulation is altered in Type I (insulin-dependent) diabetes mellitus. *Diabetologia* 1984; 27:509-513.

Malone JI, Leavengood H, Peterson MJ, et al: Red blood cell sorbitol as an indicator of polyol pathway activity. *Diabetes* 1984; 33:45-49.

Maragoudakis ME, Kelinsky H, Wasvary J, et al: Inhibition of basement membrane synthesis and aldose reductase activity by GPA 1734. *Fed Proc* 1976; 35:679.

Maragoudakis ME, Wasvary J, Gaigiulo P, et al: Human placental aldose reductase: Sensitive and insensitive forms to inhibition by alrestatin. *Fed Proc* 1979; 30:255(A).

Markus HB, Raducha M, Harris H: Tissue distribution of mammalian aldose reductase and related enzymes. *Biochem Med* 1983; 29:31–45.

Martines-Hernandez A, Amenta P: The basement membrane in pathology. *Lab Invest* 1983; 48:656–677.

Martyn CN, Matthews DM, Popp-Snijders C, et al: Effects of Sorbinil treatment on erythrocytes and platelets of persons with diabetes. *Diabetes Care* 1985; 9:36–39.

Mayer JH, Herberg L, Tomlinson DR: Axonal transport and nerve conduction and their relation to nerve polyol and *myo*-inositol levels in spontaneously diabetic BB/D rats. *Neurochem Pathol* 1984; 2:285–293.

Mayer JH, Tomlinson DR: Prevention of defects of axonal transport and nerve conduction velocity by oral administration of *myo*-inositol or an aldose reductase inhibitor in streptozotocin-diabetic rats. *Diabetologia* 1983; 25:433–438.

Mayer JH, Tomlinson DR: Axonal transport of cholinergic transmitter enzymes in vagus and sciatic nerves of rats with acute experimental diabetes mellitus; correlation with motor nerve conduction velocity and effects of insulin. *Neuroscience* 1983; 9:951–957.

Mayer JH, Tomlinson Dr, McLean WG: Slow orthograde axonal transport of radiolabelled protein in sciatic motoneurones of rats with short-term experimental diabetes: Effects of treatment with an aldose reductase inhibitor or *myo*-inositol. *J Neurochem* 1984; 43:1265–1270.

Mayhew JA, Gillon KRW, Hawthorne JN: Free and lipid inositol and sugars in sciatic nerve obtained postmortem from diabetic patients and control subjects. *Diabetologia* 1983; 24:13–15.

Medori R, Autilio-Gambetti L, Monaco S, et al: Experimental diabetic neuropathy: Impairment of slow transport with changes in axon cross-sectional area. *Proc Natl Acad Sci USA* 1985; 82:7716–7720.

Michell RH: Inositol phospholipids and cell surface receptor function. *Biochim Biophys Acta* 1980; 415:81–147.

Monckton G, Pehoevich E: Autonomic neuropathy in the streptozotocin diabetic rat. *Can J Neurol Sci* 1980; 7:135–142.

Moonsammy GI, Stewart MA: Purification and properties of brain aldose reductase and L-hexonate dehydrogenase. *J Neurochem* 1967; 14:1187–1193.

Moore SA, Peterson RG, Felton DL, et al: Reduced sensory and motor conduction velocity in 25-week-old diabetic [C57BL/K$_s$ (db/db)] mice. *Exp Neurol* 1980; 70:548–555.

Morrison AD, Clements RS Jr, Winegrad AI: Effects of elevated glucose concentrations on the metabolism of the aortic wall. *J Clin Invest* 1972; 51:3114–3123.

Morrison AD, Clements RS Jr, Travis SB, et al: Glucose utilization by the polyol pathway in human erythrocytes. *Biochem Biophys Res Commun* 1970; 40:199–205.

Myers RR, Costello ML, Powell HC: Increased endoneural fluid pressure in galactose neuropathy. *Muscle Nerve* 1979; 2:229–303.

Morrison AD: Linkage of polyol pathway activity and *myo*-inositol in aortic smooth muscle. *Diabetes* 1985; 34(suppl 1):12A.

Natarajan V, Dyck PJ, Schmid HO: Alterations in inositol lipid metabolism of rat sciatic nerve in streptozotocin-induced diabetes. *J Neurochem* 1981; 36:413–419.

Nishizuka Y: Turnover of inositol phospholipids and signal transduction. *Science* 1984; 225:1365–1370.

Obazawa H, Merola LO, Kinoshita JH: The effects of xylose on the isolated lens. *Invest Ophthalmol Vis Sci* 1974; 13:204–209.

O'Brien MM, Schofield PJ: Polyol pathway enzymes of human brain. *Biochem J* 1980; 187:21–30.

O'Brien MM, Schofield PJ, Edwards MR: Inhibition of human brain aldose reductase and hexonate dehydrogenase by alrestatin and Sorbinil. *J Neurochem* 1982; 39:810–814.

Ohrloff C, Zierz S, Hockwin O: Investigations of the enzymes involved in the fructose breakdown in the cattle lens. *Ophthalmic Res* 1982; 14:221–229.

Okuda J, Miwa I, Inagaki K, et al: Inhibition of aldose reductases from rat and bovine lenses by flavinoids. *Biochem Pharmacol* 1982; 31:3807–3822.

Ono H, Hayano S: 2,2′,4′4′-Tetrahydroxybenzophenone as a new aldose reductase inhibitor. *Nippon Ganka Gakkai Zasshi* 1982; 86:353–357.

Ono H, Nozawa Y, Hayano S: Effects of M-79,175, an aldose reductase inhibitor, on experimental sugar cataracts. *Nippon Ganka Gakkai Zasshi* 1982; 86:1343–1350.

Palamano KP, Whiting PH, Hawthorne JN: Free and lipid *myo*-inositol in tissues from rats with acute and less severe streptozotocin-induced diabetes. *Biochem J* 1977; 167:229–235.

Parathasarathy N, Spiro RG: Effect of diabetes on the glycosaminoglycan component of the human glomerular basement membrane. *Diabetes* 1982; 31:738–741.

Perejda AJ, Uitto J: Nonenzymatic glycosylation of collagen and other proteins: Relationship to the development of diabetic complications. *Coll Rel Res* 1982; 2:81–88.

Perry HD, Foulks GN, Thoft RA, et al: Corneal complications after closed vitrectomy through the pars plana. *Arch Ophthalmol* 1978; 96:1401–1403.

Peterson MJ, Sarges R, Aldinger CE, et al: CP-45,634: A novel aldose reductase inhibitor that inhibits polyol pathway activity in diabetic and galactosemic rats. *Metabolism* 1979; 28(suppl 1):456–461.

Peterson MJ, Sarges R, Aldinger CE, et al: Inhibition of polyol pathway activity in diabetic and galactosemic rats by the aldose reductase inhibitor CP-45,634. *Adv Exp Med Biol* 1979; 119:347–356.

Pfeifer MA, Weinberg CR, Cook DL, et al: Correlations among autonomic, sensory and motor neural function tests in untreated non-insulin-dependent diabetic individuals. *Diabetes Care* 1985; 8:576–584.

Pfister RR, Schepens CL, Lemp MA, et al: Photocoagulation keratopathy: Report of a case. *Arch Ophthalmol* 1971; 86:94–96.

Pfister JR, Wymann WE, Mahoney JM, et al: Synthesis and aldose reductase inhibitor activity of 7-sulfamoylxanthone-2-carboxylic acids. *J Med Chem* 1980; 23:1264–1267.

Pitkänen E: The serum polyol pattern and the urinary polyol excretion in diabetic and uremic patients. *Clin Chim Acta* 1972; 38:221–230.

Pitkänen E, Servo C: Cerebrospinal fluid polyols in patients with diabetes. *Clin Chim Acta* 1973; 44:437–442.

Popp-Snijders C, Lomecky-Janousek MZ, Schouten J, et al: *Myo*-inositol and sorbitol in erythrocytes from diabetic patients before and after Sorbinil treatment. *Diabetologia* 1984; 27:514–516.

Poulsom R, Boot-Hanford RP, Heath H: Some effects of aldose reductase inhibition upon the eyes of long-term streptozotocin-diabetic rats. *Curr Eye Res* 1982; 2:351–355.

Poulsom R, Heath H: Inhibition of aldose reductase in five tissues of the streptozotocin diabetic rat. *Biochem Pharmacol* 1983; 32:1495–1499.

Poulsom R, Mirrlees DJ, Earl DCN, et al: The effects of an aldose reductase inhibitor upon the sorbitol pathway, fructose-1-phosphate and lactate in the retina and nerve of streptozotocin diabetic rats. *Exp Eye Res* 1983; 36:751–760.

Powell HC: Pathology of diabetic neuropathy: New observations, new hypotheses. *Lab Invest* 1983; 49:515–518.

Powell HC, Costello ML, Myers RR: Endoneurial fluid pressure in experimental models of diabetic neuropathy. *J Neuropathol Exp Neurol* 1981; 40:613–634.

Powell H, Know D, Lee S, et al: Alloxan diabetic neuropathy: Electron microscopic studies. *Neurology* 1977; 27:60–66.

Powell HC, Myers RR: Schwann cell changes and demyelination in chronic galactose neuropathy. *Muscle Nerve* 1983; 6:218–227.

Pozza G, Cordaro C, Carandente O, et al: Study on relationship between erythrocyte filtration and other risk factors in diabetic angiopathy. *Ric Clin Lab* 1981; 11(suppl 1):317–326.

Proceedings of Puerto Rico Conference on Sorbinil and Diabetic Complications. *Metabolism* 35(suppl 1), April 1986.

Proceedings of Aldose Reductase Inhibitor Symposium, London, 1984. *Diabetic Medicine*, vol 2, 1985.

Puhakainen E, Saamanen AM, Lehtinen J, et al: The effect of aldose reductase inhibitor (Sorbinil®) on erythrocyte sorbitol concentration in diabetic neuropathy. *Acta Endocrinol* 1983; 103(suppl 257):56.

Raff MC, Asbury AK: Ischemic mononeuropathy and mononeuropathy multiplex in diabetes mellitus. *N Engl J Med* 1968; 279:17–22.

Raskin P, Rosenstock J, Challis P, et al: Effect of tolrestat on RBC sorbitol levels in diabetic subjects. *Diabetes* 1985; 34(suppl 1):7A.

Reddy DVN: Amino acid transport in the lens in relation to sugar cataracts. *Invest Ophthalmol* 1965; 4:700.

Reddy VN: Metabolism of glutathione in the lens. *Exp Eye Res* 1971; 11:310–328.

Reddy VN, Chakrapani B, Steen D: Sorbitol pathway in the ciliary body in relation to accumulation of amino acids in the aqueous humor of alloxan diabetic rabbits. *Invest Ophthalmol* 1971; 100:870–875.

Reddy DVN, Kinsey VE: Transport of amino acids into intraocular fluids and lens in diabetic rabbits. *Invest Ophthalmol* 1963; 2:237–242.

Reddy VN, Schauss D, Chakrapani B, et al: Biochemical changes associated with the development and reversal of galactose cataracts. *Exp Eye Res* 1976; 23:483–493.

Richon AB, Maragoudakis ME, Wasvary JS: Isoxazolidine-3,5-diones as lens aldose reductase inhibitors. *J Med Chem* 1982; 25:745–747.

Ris MM, von Wartbury J: Heterogeneity of NADPH-dependent aldehyde reductase from human and rat brain. *Eur J Biochem* 1973; 37:69–77.

Ristelli J, Koivisto VA, Akerblom HK, et al: Intracellular enzymes of collagen biosynthesis in rat kidney with streptozotocin diabetes. *Diabetes* 1976; 25:1066–1070.

Robey C, Dasmahapatra A, Cohen MP, et al: Sorbinil prevents decreased erythrocyte deformability in diabetes. *Diabetes* 1985; 34:161A.

Robey C, Dasmahapatra A, Cohen MP: In vitro effects of hyperglycemia and Sorbinil on erythrocyte (RBC) deformability. *Diabetes* 1986; 35:108A.

Robey C, Dasmahapatra A, Cohen MP, et al: Sorbinil prevents decreased erythrocyte deformability in diabetes mellitus. Manuscript submitted, 1986.

Robison WG, Kador PF, Kinoshita JH: Retinal capillaries: Basement membrane thickening by galactosemia prevented with aldose reductase inhibitor. *Science* 1983; 221:1177–1179.

Roelofsen B, Trip MVL-S: The fraction of phosphatidylinositol that activates the (Na^+/K^+)-ATPase in rabbit kidney microsomes is closely associated with the enzyme protein. *Biochim Biophys Acta* 1981; 647:302–306.

Rohrbach DH, Hassell JR, Kleinman HK, et al: Alterations in the basement membrane (heparan sulfate) proteoglycan in diabetic mice. *Diabetes* 1982; 31:185–188.

Rohrbach DH, Wagner CW, Star VL, et al: Reduced synthesis of basement membrane heparan sulfate proteoglycan in streptozotocin-induced diabetic mice. *J Biol Chem* 1983; 258:11672–11677.

Romen W, Heck T, Rauscher G, et al: Glomerular basement membrane turnover in young, old, and streptozotocin-diabetic rats. *Renal Physiol* 1980; 3:324–329.

Romen W, Lange H-W, Hempel K, et al: Studies on collagen metabolism in rats. II. Turnover and amino acid composition of the collagen of glomerular basement membrane in diabetes mellitus. *Virchows Arch (Cell Pathol)* 1981; 36:313–320.

Ross WM, Creighton MO, Stewart-DeHaan PJ, et al: Modelling cortical cataractogenesis: 3. In vivo effects of vitamin E on cataractogenesis in diabetic rats. *Can J Ophthalmol* 1982; 17:61–66.

Russell P, Merola LO, Yajima Y, et al: Aldose reductase activity in a cultured human retinal cell line. *Exp Eye Res* 1982; 35:331–336.

Satoh M, Imaizumi K, Bessho T, et al: Increased erythrocyte aggregation in diabetes and its relationship to glycosylated hemoglobin and retinopathy. *Diabetologia* 1984; 27:517–521.

Schlaepfer WW, Gerritsen GC, Dulin WE: Segmental demyelination in the distal peripheral nerves of chronically diabetic chinese hamsters. *Diabetologia* 1974; 10:541–548.

Schmidt RE, Matschinsky FM, Godfrey DA, et al: Fast and slow axoplasmic flow in sciatic nerve of diabetic rats. *Diabetes* 1975; 24:1081–1085.

Schmid-Schonbein H, Volger E: Red cell aggregation and red cell deformability in diabetes. *Diabetes* 1976; 25:897–902.

Schnur RC, Sarges R, Peterson MJ: Spiro oxazolidinedione aldose reductase inhibitors. *J Med Chem* 1982; 25:1451–1454.

Schultz RO, VanHorn DL, Peters MA, et al: Diabetic keratopathy. *Trans Am Ophthalmol Soc* 1981; 79:180–199.

Segelman AB, Segelman FP, Varma SD, et al: *Cannabis sativa L.* (Marijuana) IX: Lens aldose reductase inhibitory activity of marijuana flavone C-glycosides. *J Pharm Sci* 1977; 66:1358–1359.

Seneviratne KN: Permeability of blood nerve barriers in the diabetic rat. *J Neurol Neurosurg Psychiatry* 1972; 35:156–162.

Servo C: Sorbitol and *myo*-inositol in the cerebrospinal fluid of diabetic patients. *Acta Endocrinol* 1980; 94(suppl 238):133–138.

Servo C, Bergstrom L, Fogelholm R: Cerebrospinal fluid sorbitol and *myo*-inositol in diabetic polyneuropathy. *Acta Med Scand* 1977; 202:301–304.

Servo C, Pitkänen E: Variation in polyol levels in cerebrospinal fluid and serum in diabetic patients. *Diabetologia* 1975; 11:575–580.

Seyer-Hansen K: Renal hypertrophy in experimental diabetes: Relation to severity of diabetes. *Diabetologia* 1977; 13:141–143.

Sestanj K, Bellini F, Fung S, et al: *N*([5-(trifluoromethyl)-6-methoxy-1-naphthalenyl] thioxomethyl)-*N*-methylglycine (tolrestat), a potent orally active aldose reductase inhibitor. *J Med Chem* 1984; 27:255–256.

Shakib M, Cunha-Vaz JG: Studies on the permeability of the blood retinal barrier. IV. Junctional complexes of the retinal vessels and their role in the permeability of the blood retinal barrier. *Exp Eye Res* 1966; 5:229–234.

Sharma AK, Baker RWR, Thomas PK: Peripheral nerve abnormalities related to galactose administration in rats. *J Neurol Neuropathol Psychiatry* 1976; 39:794–802.

Sharma YR, Cotlier E: Inhibition of lens and cataract aldose reductase by protein-bound anti-rheumatic drugs: Salicylate, indomethacin, oxyphenbutazone, sulindac. *Exp Eye Res* 1982; 35:21–27.

Sheaf CM, Doughty CC: Physical and kinetic properties of homogenous bovine lens. *J Biol Chem* 1976; 251:2696–2702.

Sheys GH, Arnold WJ, Watson JA, et al: Aldose reductase from *Rhodotorula*. *J Biol Chem* 1971; 246:3824–3827.

Sheys GH, Doughty CC: The reaction mechanism of aldose reductase from *Rhodotorula*. *Biochim Biophys Acta* 1971; 242:523–531.

Sherman WR, Stewart MA: Identification of sorbitol in mammalian nerve. *Biochem Biophys Res Commun* 1966; 22:492–497.

Sidenius P: The axonapathy of diabetic neuropathy. *Diabetes* 1982; 31:356–363.

Siddiqui MA, Rahman MA: Effect of hyperglycemia on the enzyme activities of lenticular tissue of rats. *Exp Eye Res* 1980; 31:463–469.

Sima AA: Peripheral neuropathy in the spontaneously diabetic BB-Wistar rat. *Acta Neuropathol (Berl)* 1980; 51:223–227.

Sima AAF, Bril V, Greene DA: A new characteristic ultrastructural abnormality, and morphologic evidence for pathogenetic heterogeneity in human diabetic neuropathy. *Clin Res* 1986; 34:688A.

Sima AAF, Lattimer SA, Yagihashi S, et al: Biochemical, functional, and structural correction of diabetic neuropathy in the BB-rat after insulin treatment. *Fed Proc* 1984; 43:375A.

Sima AF, Lattimer SA, Yagihashi S, et al: Axo-glial dysfunction. A novel structural lesion that accounts for poorly reversible slowing of nerve conduction in the spontaneously diabetic Bio-Breeding rat. *J Clin Invest* 1986; 77:474–484.

Simard-Duquesne N, Greslin E, Dubuc J, et al: The effect of a new aldose reductase inhibitor (tolrestat) in galactosemic and diabetic rats. *Metabolism* 1985; 34:885–892.

Simard-Duquesne N, Greslin E, Gonzalez R, et al: Prevention of cataract development in severely galactosemic rats by the aldose reductase inhibitor, tolrestat. *Proc Soc Exp Biol Med* 1985; 178:599–605.

Simmons DA, Kern EFO, Winegrad AI, et al: Basal phosphatidylinositol turnover controls aortic Na^+/K^+ ATPase activity. *J Clin Invest* 1986; 77:503–513.

Simmons DA, Winegrad AI, Martin DB: Significance of tissue *myo*-inositol concentrations in metabolic regulation in nerve. *Science* 1982; 217:848–851.

Sippel TO: Changes in the water, protein and glutathione contents of the lens in the course of galactose cataract development in rats. *Invest Ophthalmol* 1960; 5:568–575.

Sohda T, Mizuno K, Imamiya E, et al: Studies on antidiabetic agents. 5-Arylthiaz-olidine-2,4-diones as potent aldose reductase inhibitors. *Chem Pharm Bull (Tokyo)* 1982; 30:3601–3616.

Speiser P, Gittelsohn AM, Patz A: Studies on diabetic retinopathy. III. Influence of diabetes on intramural pericytes. *Arch Ophthalmol* 1968; 80:332–337.

Spiro RG, Spiro MJ: Effect of diabetes on the biosynthesis of renal glomerular basement membrane. Studies on the glucosyltransferase. *Diabetes* 1971; 20:641–648.

Srivastava SK, Ansari NH, Hair GA, et al: Aldose and aldehyde reductase in human tissues. *Biochim Biophys Acta* 1984; 800:220–227.

Srivastava SK, Hair GA, Das B: Activated and unactivated forms of human erythrocyte aldose reductase. *Proc Natl Acad Sci USA* 1985; 82:7222–7226.

Srivastava SK, Petrash JM, Sadana IJ, et al: Susceptibility of aldehyde and aldose reductase of human tissues to aldose reductase inhibitors. *Curr Eye Res* 1982; 2:407–410.

Stewart MA, Passoneau JV: Identification of fructose in mammalian nerve. *Biochem Biophys Res Commun* 1964; 17:536–541.

Stewart MA, Sherman WR, Anthony S: Free sugars in alloxan diabetic rat nerve. *Biochem Biophys Res Commun* 1966; 22:488–491.

Stewart MA, Sherman WR, Kurien MM, et al: Polyol accumulation in nervous tissue of rats with experimental diabetes and galactosemia. *J Neurochem* 1977; 14:1057–1066.

Stockard C, Clements R: Usefulness of blood elements in the prediction of the fructose and *myo*-inositol content of nerve and lens. *Diabetes* 1984; 33(suppl 1): 89A.

Sugimura K, Windebank AJ, Natarajan V, et al: Interstitial hyperosmolarity may cause axis cylinder shrinkage in streptozotocin diabetic nerve. *J Neuropathol Exp Neurol* 1980; 39:710–721.

Tanimoto T, Fukuda H, Kawamura J: Purification and some properties of aldose reductase from rabbit lens. *Chem Pharm Bull (Tokyo)* 1983; 31:2395–2403.

Tetjak AG, Limarenko IM, Lossova GV, et al: Interrelation of phosphoinositide metabolism and ion transport in crab nerve fiber. *J Neurochem* 1977; 28:199–205.

Thomas PK, Lascelles RG: The pathology of diabetic neuropathy. *Q J Med* 1966; 24:489–509.

Thornally PJ, Wolff SP, Crabbe MJC, et al: Auto-oxidation of glyceraldehyde and other monosaccharides catalyzed by buffer ions. *Biochim Biophys Acta* 1984; 797:276–287.

Timperley WR: Vascular and coagulation abnormalities in diabetic neuropathy and encephalopathy. *Horm Metab Res* 1980; 9(suppl):43–49.

Timperley WR, Ward JD, Preston FE, et al: Clinical and histological studies in diabetic neuropathy: A reassessment of vascular factors in relation to vascular coagulation. *Diabetologia* 1976; 12:237–243.

Toback FG: Phosphatidylcholine metabolism during renal growth and regeneration. *Am J Physiol* 1984; 246:F249–F259.

Tomlinson DR, Holmes PR, Mayer JH: Reversal, by treatment with an aldose reductase inhibitor, of impaired axonal transport and motor nerve conduction velocity in experimental diabetes. *Neurosci Lett* 1982; 31:189–193.

Tomlinson DR, Mayer JH: Defects of axonal transport in diabetes mellitus—a possible contribution to the aetiology of diabetic neuropathy. *J Auton Pharmacol* 1984; 4:59–72.

Tomlinson DR, Moriarty RJ, Mayer JH: Prevention and reversal of defective axonal transport and motor nerve conduction velocity in rats with experimental diabetes by treatment with the aldose reductase inhibitor Sorbinil. *Diabetes* 1984; 33:470–476.

Tomlinson DR, Sidenius P, Larsen JR: Slow component-a of axonal transport, nerve *myo*-inositol, and aldose reductase inhibition in diabetic rats. *Diabetes* 1986; 35:398–402.

Tomlinson DR, Townsend J: Protection by the aldose reductase inhibitor "Statil" (ICI 128436) against loss of an axonally transported enzyme in nerve terminals of diabetic rats. *Diabetes* 1985; 34(suppl 1):202A.

Tomlinson DR, Townsend J, Fretten P: Prevention of defective axonal transport in streptozotocin-diabetic rats by treatment with "Statil" (ICI 128436), an aldose reductase inhibitor. *Diabetes* 1985; 34:970–972.

Travis SF, Morrison AD, Clements RS Jr, et al: Metabolic alterations in the human erythrocyte produced by increases in glucose concentration. *J Clin Invest* 1971; 50:2104–2112.

Trüeb B, Fluckiger R, Winterhalter KH: Nonenzymatic glycosylation of basement membrane collagen in diabetes mellitus. *Coll Rel Res* 1984; 4:239–251.

Turner AJ, Tipton KF: The characterization of two reduced nicotinamide-adenine dinucleotide phosphate-linked aldehyde reductases from pig brain. *Biochem J* 1972; 130:765–772.

Uitto J, Perejda AJ, Grant GA, et al: Glucosylation of human glomerular basement membrane collagen: Increased content of hexose in ketoamine linkage and unaltered hydroxylysine-o-glycosides in patients with diabetes. *Connect Tiss Res* 1982; 10:287–296.

Ulbrecht JS, Bensen JT, Greene DA: Sodium-dependent *myo*-inositol uptake in isolated renal glomeruli: Possible inhibition by glucose. *Diabetes* 1985; 34 (suppl 1):13A.

Unakar NJ, Tsui JY: Inhibition of galactose-induced alterations in ocular lens with Sorbinil. *Exp Eye Res* 1983; 36:685–694.

Valbo AB, Hagbarth K-E, Torebjork HE, et al: Somatosensory, proprioceptive and sympathetic activity in human peripheral nerves. *Physiol Rev* 1979; 59:919–957.

Van Heyningen R: Formation of polyols by the lens of the rat with sugar cataracts. *Nature (Lond)* 1959; 184:194–195.

Van Heyningen R: Metabolism of xylose by the lens. Rat lens in vivo and in vitro. *Biochem J* 1959; 73:197–207.

Varma SD: Superoxide and lens of the eye. A new theory of cataractogenesis. *Int J Quantum Chem* 1981; 20:479–484.

Varma SD, El-Aguizy HK, Richards RD: Refractive changes in alloxan diabetic rabbits. Control by flavinoids. *Acta Ophthalmol* 1980; 58:748–759.

Varma SD, Kinoshita JH: The absence of cataracts in mice with congenital hyperglycemia. *Exp Eye Res* 1974; 19:577–582.

Varma SD, Kinoshita JH: Sorbitol pathway in diabetic galactosemic rat lens. *Biochim Biophys Acta* 1974; 338:632–640.

Varma SD, Kinoshita JH: Inhibition of lens aldose reductase by flavinoids. *Biochem Pharmacol* 1976; 25:2505–2513.

Varma SD, Kumar S, Richards RD: Protection by ascorbate against superoxide injury to the lens. *Invest Ophthalmol Vis Sci* 1979; 18(suppl):98.

Varma SD, Mikuni I, Kinoshita JH: Flavinoids as inhibitors of lens aldose reductase. *Science* 1975; 188:1215–1216.

Varma SD, Mizuno A, Kinoshita JH: Diabetic cataracts and flavinoids. *Science* 1977; 195:205–206.

Varma SD, Shockert SS, Richards RD: Implications of aldose reductase in cataracts in human diabetics. *Invest Ophthalmol Vis Sci* 1979; 18:237–241.

Vere DW, Verrel D: Relation between blood sugar level and the optical properties of the lens of the human eye. *Clin Sci* 1955; 14:183–196.

Vogt BW, Schleicher ED, Wieland OH: ϵ-Amino-lysine bound glucose in human tissues obtained at autopsy: Increase in diabetes mellitus. *Diabetes* 1983; 31:1123–1127.

Waltman S, Krupin T, Hanish S, et al: Alteration of the blood retinal barrier in experimental diabetes mellitus. *Arch Ophthalmol* 1978; 96:878–879.

Ward JD: The polyol pathway in the neuropathy of early diabetes, in Camerini-Davalos RA, Cole HS (eds): *Vascular and Neurologic Changes in Early Diabetes*, pp 525–532. New York, Academic Press, 1973.

Ward JD, Baker RWP, Davis BH: Effect of blood sugar control on the accumulation of sorbitol and fructose in nerve tissues. *Diabetes* 1972; 21:1173–1178.

Wautier JL, Paton RC, Wautier M-P, et al: Increased adhesion of erythrocytes to endothelial cells in diabetes mellitus and its relation to vascular complications. *N Engl J Med* 1981; 305:237–242.

Wells AM, Greene DA: A Sorbinil-responsive *myo*-inositol related Na/K-ATPase defect in diabetic rat superior cervical ganglion. *Diabetes* 1985; 34(suppl 1): 102A.

Wermuth B, Burgesser H, Bohren K, et al: Purification and characterization of human-brain aldose reductase. *Eur J Biochem* 1982; 127:297–284.

White GL, Larabee MG: Phosphoinositides and other phospholipids in sympathetic ganglia and nerve trunks of rats. *J Neurochem* 1973; 20:783–798.

White GL, Schellhase HU, Hawthorne JN: Phosphoinositide metabolism in rat superior cervical ganglion, vagus and phrenic nerve: Effects of electrical stimulation and various blocking agents. *J Neurochem* 1974; 28:149–158.

Whiting PH, Palmano KP, Hawthorne JN: Enzymes of *myo*-inositol and inositol lipid metabolism in rats with streptozotocin induced diabetes. *Biochem J* 1979; 179:549–553.

Williams E, Timperley WR, Ward JD, et al: Electron microscopic studies of vessels in diabetic peripheral neuropathy. *J Clin Pathol* 1980; 33:462–470.

Wick AN, Drury DR: Insulin and permeability of cells to sorbitol. *Am J Physiol* 1951; 166:421–423.

Williamson JR, Chang K, Rowold E, et al: Diabetes-induced increases in vascular permeability are prevented by castration and by Sorbinil. *Diabetes* 1985; 34(suppl 1):108A.

Williamson JR, Chang C, Rowold E, et al: Diabetes-induced increases in vascular permeability and changes in granulation tissue levels of sorbitol, *myo*-inositol, *chiro*-inositol, and *scyllo*-inositol are prevented by Sorbinil. *Metabolism* 1986; 35(suppl 1):41–45.

Williamson JR, Chang K, Rowold E, et al: Sorbinil prevents diabetes-induced increases in vascular permeability but does not alter collagen cross-linking. *Diabetes* 1985; 34:703–705.

Williamson JR, Rowold E, Chang K, et al: Albumin permeation of new (granulation tissue) vessels is increased in diabetic rats. *Diabetes* 1984; 33(suppl 1):3A.

Williamson JR, Rowold E, Chang K, et al: Sex steroid dependency of diabetes-induced changes in polyol metabolism, vascular permeability, and collagen cross-linking. *Diabetes* 1986; 35:20–27.

Winegrad AI, Greene DA: Diabetic polyneuropathy: The importance of insulin deficiency, hyperglycemia and alterations in *myo*-inositol metabolism in its pathogenesis. *N Engl J Med* 1976; 295:1416–1421.

Winegrad AI, Simmons DA, Martin DB: Has one diabetic complication been explained? *N Engl J Med* 1983; 308:152–154.

Wirth H-P, Wermuth B: Immunochemical characterization of aldo-keto reductases from human tissues. *FEBS Lett* 1985; 187:280–282.

Wise GN, Dollery GT, Henkind P: *The Retinal Circulation*, pp 290–324, 350–420. New York, Harper & Row, 1971.

Wolff SP, Crabbe MJC, Thornally PJ: Autoxidation of glyceraldehyde and other simple monosaccharides. *Experientia* 1984; 40:244–246.

Wu V-Y, Cohen MP: Platelet factor 4 binding to glomerular microvascular matrix. *Biochim Biophys Acta* 1984; 797:76–82.

Yagihashi S, Kudo K, Nishihira M: Peripheral nerve structures of experimental diabetes in rats and the effect of insulin treatment. *Tohuku J Exp Med* 1979; 127:35–44.

Yamani T: Effect of aldose reductase inhibitor on the oscillatory potential in ERG of streptozotocin diabetic rats. *Folia Ophthalmol Jpn* 1983; 34:2237–2244.

Yanoff M: Diabetic retinopathy. *N Engl J Med* 1966; 274:1344–1349.

Yoshida H: The characteristics of aldose reductase in human lens, placenta, and rat organs. *Nippon Ganka Gakkai Zasshi* 1981; 85:865–869.

Young RJ, Ewing DJ, Clarke BF: A controlled trial of Sorbinil, an aldose reductase inhibitor, in chronic painful diabetic neuropathy. *Diabetes* 1983; 32:938–942.

Young RJ, Matthews DM, Clarke BF, et al: Aldose reductase inhibition for diabetic neuropathy (letter to the editor). *Lancet* 1983; 2:969.

Yue DK, Hanwell MA, Satchell PM, et al: The effect of aldose reductase inhibition on motor nerve conduction velocity in diabetic rats. *Diabetes* 1982; 31:789–794.

Index